Springer Series on Atomic, Optical, and Plasma Physics

Volume 98

The Springer Series on Atomic, Optical, and Plasma Physics covers in a comprehensive manner theory and experiment in the entire field of atoms and molecules and their interaction with electromagnetic radiation. Books in the series provide a rich source of new ideas and techniques with wide applications in fields such as chemistry, materials science, astrophysics, surface science, plasma technology, advanced optics, aeronomy, and engineering. Laser physics is a particular connecting theme that has provided much of the continuing impetus for new developments in the field, such as quantum computation and Bose-Einstein condensation. The purpose of the series is to cover the gap between standard undergraduate textbooks and the research literature with emphasis on the fundamental ideas, methods, techniques, and results in the field.

More information about this series at http://www.springer.com/series/411

Inga Tolstikhina · Makoto Imai
Nicolas Winckler · Viacheslav Shevelko

Basic Atomic Interactions of Accelerated Heavy Ions in Matter

Atomic Interactions of Heavy Ions

 Springer

Inga Tolstikhina
Lebedev Physical Institute
Moscow
Russia

Makoto Imai
Department of Nuclear Engineering
Kyoto University
Kyoto
Japan

Nicolas Winckler
GSI Helmholtzzentrum für
　Schwerionenforschung GmbH
Darmstadt
Germany

Viacheslav Shevelko
Lebedev Physical Institute
Moscow
Russia

ISSN 1615-5653　　　　　　ISSN 2197-6791　(electronic)
Springer Series on Atomic, Optical, and Plasma Physics
ISBN 978-3-030-09122-4　　　ISBN 978-3-319-74992-1　(eBook)
https://doi.org/10.1007/978-3-319-74992-1

Printed on acid-free paper

This Springer imprint is published by Springer Nature
The registered company is Springer International Publishing AG
The registered company address is: Gewerbestrasse 11, 6330 Cham, Switzerland

To our parents:

Galina and Yuriy Loginov,
Sachie and Junso Imai,
Marie Noelle Strecker and
Georges Winckler, and
Zinaida Tatarinova and Pyotr Shevelko

Preface

The aim of this book is to present researchers and graduated students, working in the field of atomic, plasma, and accelerator physics, the basic aspects of atomic interaction processes, occurring in penetration of fast heavy-ion beams through matter—gaseous, solid, and plasma targets. Although an interest to investigate interactions of ion beams with penetrated media has arisen more than 100 years ago, this topic is still actual because of fast development of acceleration techniques of heavy ions, progress in astrophysics, which includes unknown phenomena, as well as new material production, progress in cancer/tumor therapy, industrial tokamak devices, and plasma technique applied for effective production of microchips and integrated circuits.

This book is a continuation of a previous series of the books such as Physics of Highly Charged Ions by R. K. Janev, L. P. Presnyakov, and V. P. Shevelko, Springer, Berlin (1985), Introduction to the Physics of Highly Charged Ions by H. Beyer, V. P. Shevelko, IOP, Bristol (2003), The Physics of Multiply and Highly Charged Ions by F. J. Currell, ed., Kluwer Academic Pub., Dordrecht, Boston, London (2003), Atomic Processes in Basic and Applied Physics by V. P. Shevelko, H. Tawara, eds., Springer, Heidelberg (2012).

In the present book, the interaction processes of accelerated ions with the target particles are described and interpreted in terms of the atomic radiative and collision properties such as electron-loss and electron-capture cross sections, transition probabilities, and other characteristics. The main attention is paid to many-electron projectiles and heavy target particles. The principle peculiarity of such heavy systems is the role of inner-shell electrons which in many cases play a major role, and a contribution of the outermost electrons can be neglected. The influence of inner-shell electrons leads to a change in the scaling laws for the collision cross sections as a function of atomic parameters.

A big part of the book is devoted to consideration of the cross sections responsible for interaction of ion beams with gaseous, solid, and plasma targets, including experimental results, theoretical methods, and computer codes used for calculations such as CTMC, CAPTURE, DEPOSIT, RICODE, and others. Atomic processes are mainly responsible for kinetic energy losses of ion beams in penetrating matter (stopping power) and evolution of the charge-state fractions.

Dynamics of the charge-state fractions and equilibrium mean charges of ion beams interacting with media are considered on the basis of the balance equilibrium equations, including equilibration of the charge-state fractions and mean charges, equilibrium target thickness. A short description of the computer programs ETACHA, GLOBAL, CHARGE, BREIT, and others for calculation of the charge-state fractions as a function of the target thickness is given as well as some applications such as detection of super-heavy elements and creating an inverse population in a capillary discharge plasma.

A special aspect, considered here, is electron-capture processes at *low* collision energies and related topics such as the influence of the isotope effect on the cross sections of resonant and quasi-resonant electron capture in collisions with hydrogen isotopes H, D, and T. These processes are of high interest for specialists studying DT plasmas due to two main reasons. First, these processes exhibit the dominant mechanisms for creating in a plasma the impurity ions in excited states, radiative short-wavelength spectra of which are used for plasma diagnostics. Second, a high interest is now related to $W(Z = 74)$ atoms and ions because tungsten is considered as the most perspective element for making walls and diverter in plasma devices with magnetic confinement, where the interactions of W atoms and ions with hydrogen and its isotopes play an important role. It should be noted that the common scaling laws for electron-capture cross sections are not valid for low-energy collisions with hydrogen isotopes, and that the influence of the isotope effect on the resonant and quasi-resonant electron-capture cross sections is extremely strong: The cross sections for the reactions with different isotopes may differ by more than three orders of magnitude; therefore, investigation of these processes requires a special attention.

The book does not contain a detailed description of theoretical methods and complicated formulae but presents the data in a compact form using figures in the scaled units, tables, and simple analytical formulae which allows one to estimate the atomic characteristics (mainly cross sections) without resorting to computer. Such presentation may be of interest, especially to experimentalists working in the field of atomic, plasma, and accelerator physics.

Moscow, Russia   Inga Tolstikhina
Kyoto, Japan   Makoto Imai
Paris, France   Nicolas Winckler
                                                       Viacheslav Shevelko

Acknowledgements

The authors would like to thank many colleagues and friends around the world for their great contribution to atomic and accelerator physics included to many topics discussed in the book.

In connection with experimental investigations of fast heavy ions in gases and solid targets, we are very much obliged to the late Fritz Bosch, H. Weick, Ch. Scheidenberger, H. Geissel, Th. Stöhlker, Yu. Litvinov, Ch. E. Düllmann, J. Khuyagbaatar, A. Yakushev, and O. Rosmej for helpful and useful discussions at GSI, Darmstadt, as well as to the late Toshizo Shirai, M. Sataka, M. Matsuda, S. Okayasu, K. Kawatsura, K. Takahiro, K. Komaki, H. Shibata, K. Nishio for fruitful discussions at the tandem accelerator facility in Japan Atomic Energy Agency (JAEA), Tokai.

Experimental and theoretical works, especially on CTMC calculations, performed by R. Olson and R. D. DuBois, made a large contribution to our understanding of charge-changing processes in collisions of heavy especially uranium ions, with gaseous targets including multiple-electron-loss and electron-capture processes.

The authors appreciate very much experimental and theoretical works performed with Hiro Tawara (NIFS, Toki, Japan), whose knowledge in many fields of atomic and accelerator physics are reflected in many works and books cited in the present work. During the course of writing this book, we benefitted from a fruitful exchange of ideas and discussions with Z. Patyk, C. Rappold, M. Trassinelli, E. Lamour, D. Vernhet, J. P. Rozet, and P. Sigmund.

We have a special pleasure to thank E. Salzborn and A. Müller from Strahlenzentrum Giessen, Germany, for their fundamental experimental and theoretical results on electron–ion, ion–atom, and ion–ion collisions which have created a solid database for many branches of atomic physics.

Russian physicists V. M. Shabaev, I. I. Tupitsyn, B. M. Smirnov, V. S. Lebedev, O. I. Tolstikhin, and A. A. Narits increased our knowledge in QED, relativistic effects, and other fields of theoretical physics which helped us very much in writing and interpreting the material of the present book.

V. P. S and M. I. would like to appreciate A. Itoh and Y.-D. Jung for their continuous collaboration. M. I. would like to thank H. Brian and Carolyn Gilbody, Jörg and Midori Eichler, Robert E. H. and Sarah Clark, N. Stolterfoht and H. Moriyama for their steady encouragement as well as financial support by JSPS KAKENHI Grant Numbers 14780381, 24561024, 16K06937 and Cooperative Research Program of JAEA and University of Tokyo.

Special thanks go to the publishers, especially, to Senior Editor Claus Ascheron and Assistant Editor Adelheid Duhm for active support.

Contents

Units and Notations

Units

The system of atomic units is used unless otherwise specified: $e^2 = m = \hbar = 1$.

The values in atomic units are based on the 2014 CODATA adjustment of the values of the constants. From: Mohr P. J., Newell D. B., and Taylor B. N. CODATA recommended values of the fundamental physical constants: 2014. Rev. Mod. Phys. 88 (2016).

Quantity	Numerical Value
Length (Bohr radius)	$a_0 = 0.529\ 177\ 210\ 67(12) \times 10^{-8}$ cm
Energy (Hartree energy)	$E_h = e^2/a_0 = 2\,\mathrm{Ry} = 27.211\ 386\ 02(17)$ eV
Rydberg	$1\,\mathrm{Ry} = me^4/2\hbar^2 = 13.605\ 693\ 009(84)$ eV
Velocity	$v_0 = e^2/\hbar = 2.187\ 691\ 262\ 77(50) \times 10^8$ cm/s
Time	$\tau_0 = a_0/v_0 = \hbar^3/me^4 = 1.288\ 088\ 667\ 12(58) \times 10^{-21}$ s
Cross section	$\pi a_0^2 = 0.879\ 735\ 5419(60) \times 10^{-16}$ cm^2
Velocity of light	$c = 2.997\ 92458 \times 10^{10}$ cm/s
Fine-structure constant	$\alpha = e^2/\hbar c = 1/137.035\ 999\ 139(31)$
Bohr magneton	$\mu_B = e\hbar/2m = 5.788\ 381\ 8012(26) \times 10^{-5}$ eV T

Notations

X^{q+}	Incident ion with the charge q
q, Z_1	Incident ion charge
E	Incident ion energy
Z	Atomic number

Z_N	Nuclear charge of incident ion
Z_T	Target atomic number (nuclear charge)
I	Binding energy; ionization potential; average excitation energy
n, l	Principal and orbital quantum numbers
A	Radiative transition probability
f	Oscillator strength
σ	Cross section
Y_T	Fraction of residual gas component in accelerator
τ	Ion beam lifetime
x	Target thickness
$F_q(x)$	Nonequilibrium charge-state fraction
$F_q(\infty)$	Equilibrium charge-state fraction
x_{eq}	Equilibrium thickness
$q(x)$	Mean charge of exit ions
$\bar{q}$	Equilibrium mean charge of exit ions
υ	Projectile ion velocity
υ_e	Orbital electron velocity
Z_{eff}	Effective charge of the projectile ion
b	Impact parameter
EL	Electron loss
EC	Electron capture
CE	Charge exchange
RCE	Resonant charge exchange
DE	Target-density effect
MI	Multiple-electron ionization
DI	Direct ionization
EA	Excitation-autoionization
NRC	Non-radiative capture
MEC	Multiple-electron capture
REC	Radiative electron capture
RR	Radiative recombination
DR	Dielectronic recombination
TR	Ternary (three-body) recombination
SP, S	Stopping power
CSD	Charge-state distribution
DOS	Density of states
PFC	Plasma-facing component
NPR	Neutral particle rejector
HSCC	Hyper-spherical close coupling
END	Electron–nuclear dynamics
L_{rad}	Radiation length
SHE	Super-heavy element

Chapter 1
Introduction

Abstract Characteristics of an ion beam, penetrating through matter, are changed due to interactions with media particles: the beam energy becomes lower, energy spread and angular scattering broader etc. Both incident beam ions and target particles undergo various atomic processes, which change their radiative and collisional properties. In this chapter, macro (energy loss, angular straggling, range) and micro characteristics (charge-changing processes in gaseous and solid targets), describing collisions of ion beams with different targets, are discussed. Atomic processes between incident ions and plasma particles are discussed in Chap. 7.

1.1 Role of Atomic Processes in Penetration of Ion Beams Through Matter

Interactions of ion beams penetrating gaseous, solid and plasma targets are based on the atomic elementary processes occurring between beam ions and target particles: atoms, ions, molecules and electrons. *Nuclear* reactions, leading to a change of nucleons in the projectiles and targets, have much smaller cross sections and, therefore, are neglected in a wide energy and target-thickness ranges, or nuclear processes are considered in a special way (see Chap. 10). Here, the main attention is paid to *atomic* processes which play a key role for solving many fundamental problems in atomic, plasma and accelerator physics as well as in many applications—from particle radiation therapy to development of powerful heavy-ion accelerators, from creation of new materials to simulation of biochemical processes in living cells and others (see [1–15]).

Atomic interactions of ions with penetrating medium lead to a strong evolution of ion charge-state fractions as a function of the target thickness and are principally defined by the charge-changing processes such as electron capture and loss, recombination as well as by photo-processes—dielectronic and photorecombination, radiative electron capture etc. Information about atomic characteristics (cross sections, radiation and Auger transition rates etc.) are required to solve many

© Springer International Publishing AG 2018
I. Tolstikhina et al., *Basic Atomic Interactions of Accelerated Heavy Ions in Matter*, Springer Series on Atomic, Optical, and Plasma Physics 98, https://doi.org/10.1007/978-3-319-74992-1_1

problems of atomic physics and spectroscopy, plasma physics, quantum electronics, accelerator physics and thermonuclear fusion, as well as for reliable methods of spectroscopic and particle diagnostics of laboratory and astrophysical plasmas. Properties of interaction cross sections of ions with the target and their dependencies on ion velocity, atomic structure of colliding particles and target density (the *target-density effect*) are considered in many review papers and books (see, e.g., [1, 6, 16–43]).

In recent years, due to intensive development of accelerator techniques, the interest to study processes involving heavy many-electron ions (of the type Ar^{q+}, Xe^{q+}, Au^{q+}, W^{q+}, Bi^{q+}, U^{q+}) has significantly grown up because of their use in thermonuclear fusion [39, 44], slowing down of heavy-ion beams in matter [45], fragmentation of exotic nuclei [46], generation of extreme states of matter [47], investigation of structure of new materials [9], in astrophysics [48], in beam radiotherapy [13], in the design of the new types of accelerators and storage rings [49], and many others. Information about atomic databases on electron structure, cross sections for electron-atom, ion-atom, ion-ion processes and other characteristics can be found, e.g., on the IAEA (International Atomic Energy Agency) website [50].

Optimization of the residual-gas density and composition in an accelerator to get maximal intensity and lifetimes of ion beams with desired charge states and energy (the so-called *vacuum conditions*) is one of the most important tasks in designing the modern accelerators and storage rings. For example, this issue is a subject of detailed investigation of accelerated heavy ions within the FAIR International projects (Facility for Antiproton and Ion Research) started in 2011 at GSI (Gesellschaft für Schwerionen Forschung), Darmstadt, Germany [51]. Another example is a new NICA project [52] (Nuclotron-based Ion Collider fAcility), started in 2013 at JINR (Joint Institute for Nuclear Research), Dubna, Russia, and intended to create a collider for protons and heavy ions for investigation of super-dense matter, in which collisions of two gold-ion beams (atomic number $Z = 79$) with energy of about $10\,GeV/u$ will be substantiated.

An interest to interactions of accelerated ions with penetrating media has arisen more than 100 years ago in experimental investigations of neutralization processes in collisions with gaseous targets [53]. Experiments involving accelerated heavy many-electron ions are intensively carrying out at the world largest accelerators in JINR in Dubna, CERN (European Organization for Nuclear Research), Geneva, Switzerland, GANIL (Grand Accélérateur National d'Ions Lourds in CAen), France, UNILAC (UNIversal Linear ACcelerator) in Darmstadt, Germany, NSCL (National Superconducting Cyclotron Laboratory) and SuperHILAC (Super Heavy Ion Linear ACcelerator), USA, and HIRFL (Heavy Ion Research Facility) in Lanzhou, China. Today many modern accelerator facilities such as RIBF (Radioisotope Beam Factory) (RIBF), RIKEN, Japan [54], the future FRIB (Facility for Rare Isotope Beams), MSU, USA [55], HIFF (High Intensity heavy ion Accelerator Facility), HIRFL in Lanzhou [56], and the future FAIR (Facility for Antiproton and Ion Research), Darmstadt, [51], are planning to provide high-intensity, heavy-ion beams with energies higher than 200 MeV/u.

1.2 Targets

In reliable experiments with ion beams passing through targets, not only information about beam properties (energy-loss, angular distributions) is required, but also physical and chemical properties of the target such as chemical structure, composition, density or pressure, amorphous or crystalline structure and others. Obviously, knowledge of the target properties is very essential for interpretation of experimental results, especially, as concerns to influence of the target-density (gas-solid) effect (Sects. 2.6, 4.3 and 6.6). The target-thickness values are usually expressed in units of atoms/cm^2 or g/cm^2, or given in pressure units atoms/cm^3 (Sect. 3.1). Below the properties of gaseous, plasma and liquid targets, which are used in experiments with fast heavy-ion beams (see, e.g., [45]) are briefly described.

1.2.1 Solid Targets

The solid targets (foils) are made from metals of elements from Li ($Z = 3$) up to U ($Z = 92$) and used in practice, where Z is the atomic number of the element. The techniques used for producing foils strongly dependend on the atomic number and the target thickness. The thickness range of solid targets is very broad and varies from a few µg/cm^2 for projectiles with energy of a few keV/u, to several g/cm^2 for relativistic heavy ions [57]. The carbon foils ($Z = 6$) are mostly used since they are produced with a very high accuracy (concerning density and crystal lattice length). In recent years, *multi-layer* foils made from different materials, have started to play a relevant role in experiments to obtain a required ion charge after such multi-layer foils (see [58] and references in it).

The foils used in many experiments, for example, for detection of heavy and super-heavy elements (Sect. 9.3), should have a long lifetime, i.e., not to be destroyed and keep their atomic properties during a long irradiation of high-energy ion beam on them. This mostly concerns a thin layer of a heavy element (Pb), which really participates in a fusion reaction, deposited on a thick substrate (for example, C). The lifetime of solids, i.e., time before their demolition, depends mainly on radiative action of the incident ion beam and target evaporation, and is varying from a fraction of a second to hundreds of hours depending on the foil material, energy and charge state of the ion beam before and after the target penetration [59]. Calculations of lifetimes of foils constitute a very complicated problem and are performed with account for thermodynamic and hydrodynamic conditions which should be fulfilled in the solid targets [60, 61].

1.2.2 Gaseous Targets

As for gaseous targets, molecular hydrogen, nitrogen and noble gases are usually used in two main configurations: in special cells with solid windows to maintain high vacuum outside the cell (closed cells) and those differentially pumped with a very small aperture (windowless gas cells). The first gas targets are used to measure the stopping powers (Sect. 1.2) in gases with a relatively high density exceeding $\sim 100 \, \mu g/cm^2$ to exclude the influence of solid windows [62, 63]. The windowless gas cells are used when very thin gaseous targets are required, e.g., as an internal target in the storage-cooler rings or in experiments on measuring the cross sections where a single-collision condition is required, i.e. a low target density. The main disadvantages of such gas targets are blow-out of the target gas towards the beam path, which influences the target density and thickness estimation, and a large scattering angle that makes a detection of events more difficult. For such systems, special control methods are developed based on pulsed fast valves opened only during the beam pulse duration (see [45, 64, 65]).

1.2.3   Plasma Targets

Atomic interactions of ion beams with dense plasma targets are of importance for investigation of inertial confinement fusion driven by heavy ions, plasma diagnostics with ion beams and for obtaining maximum charge state of the exit ions. For these kinds of experiments, special plasma discharge devices are designed to be inserted in beam lines to study ion slowing down and charge-changing reactions. Such experimental setups involve an electric discharge in a quartz tube containing the gas with an initial pressure of several Torr [66, 67]. Particle abundances inside gag-filled tube depend on the gas composition, plasma density and temperature [68, 69].

Interactions of ion beams with D-T plasmas constitute a special interest [70, 71]. Determination of plasma temperature and density is usually performed by spectroscopic methods or with the laser interferometry or absorption [72, 73].

Another important direction in investigation of ion-beam-plasma interactions is a stripping in a dense plasma accompanied by an increase in the equilibrium mean charge of exit ions up to more than 10 times (and, hence, the *stopping power*) compared to the cold gas targets of the same chemical element. This mean-charge increasing takes place in the *plasma windows* at specific ion energies when the capture probability of the bound electrons in a gas target is much smaller than the probability of radiation capture of plasma free electrons [72–76].

1.2.4   Liquid Targets

Besides a pure interest to physical properties of liquids, interaction of ion beams with liquid targets is of practical interest, first of all, for biological investigations and medicine, especially, for particle cancer therapy [7, 8, 13]. Double-strand breaks of cancer-cell DNAs, brought by beams of protons and bare carbon ions at energies of a few MeV/u, are effectively used in various medical centers in Japan, USA, Germany, Russia, France, Switzerland and other countries (see [77]). Therefore, the most intensive investigations are devoted to ion beams interacting with water molecules since in a human body, water constitutes 80% of its mass. In general, in using liquid targets many physical and technical problems have to be solved related with the complicated structure of organic molecules, role of secondary electrons and chemical bonding, validity of the Bragg's additivity rule and other questions [77–80].

1.3 Characteristics of Ion Beams: Energy-Loss and Angular Straggling, Range, and Penetration Depth

1.3.1 Energy Loss

Penetration of ion beam through matter is accompanied by the beam energy loss in shifting the initial energy value to lower energies and broadening of the energy profile (see Fig. 1.1, left). Mean energy loss $<\Delta E>$ of the fast ion beam, averaged over all collisions, is given by collision fluctuations due to two main processes: ionization of the target particles and variations of the charge states as a function of the penetration depth of the beam in matter. The quantity $<\Delta E>$ is defined by [14]:

$$<\Delta E> = Nx \sum_i T_i \sigma_i = Nx \int T d\sigma = Nx \int T \frac{d\sigma}{dT} dT, \qquad (1.1)$$

where N denotes the density of the target particles, x the target thickness, T the energy loss in a single collision and σ the energy-loss cross section (see [14, 32]). Brackets mean statistic average over all collisions of beam ions with the media particles. At relatively low energies $E \approx 10\text{--}100\,\text{keV/u}$, the energy losses are defined mainly by elastic collisions with the nuclei of the media particles, and at $E \approx 10\text{--}100\,\text{MeV/u}$ they are caused by inelastic processes involving projectile and target electrons: electron capture, loss and ionization of the media particles (see Sect. 1.4).

In practice, the kinetic energy losses of the ion beam in penetrated media are characterized by the *stopping power* $S = -dE/dx$:

$$<\Delta E> = - \int_0^L \frac{dE}{dx} dx, \qquad (1.2)$$

where L denotes the target thickness.

In a *dense* matter, energy shape of the ion beam, distributed on energy due to many-fold collisions, is close to the Gaussian function with the width Ω, called *energy-loss straggling* [14]:

$$\Omega^2 = \langle(\langle\Delta E\rangle - \Delta E)^2\rangle = Nx \sum_i T_i^2 \sigma_i = Nx \int T^2 d\sigma = Nx \int T^2 \frac{d\sigma}{dT} dT.$$

(1.3)

In general, to present the Ω quantity in a close analytical form is not possible but in the case of two charge-state fractions F_q with the charge-state q_0 and q_1, the width Ω has the form [82]:

$$\Omega^2 = 2L \frac{F_0^\infty F_0^\infty}{N(\sigma_{01} + \sigma_{10})} \left[\frac{dE}{dx}(q_0) - \frac{dE}{dx}(q_1)\right]^2,$$

(1.4)

where L and N denote the thickness and density of the target, σ_{01} and σ_{10} electron loss and capture cross sections between charge states q_0 and q_1, respectively, $-dE/dx(q)$ partial stopping powers [45] and $F_{0,1}^\infty$ *equilibrium fractions* (Sect. 3.3).

In a *low* density media (dilute gas), energy shape of the ion beam is much broader than the Gaussian distribution and is described by the Landau-Vavilov formula [83, 84]. A spread of energy loss at relativistic energies is considered in [85].

Exact energy distribution of exit ions is required for many applications, for example, in production of electronic chips and semiconductor detectors, in medicine in cancer therapy where it is necessary to know ion energy before and after interaction of ion beam with a patient body. Experimental methods of deposited energy of ion beams with keV/u–MeV/u energy range are considered in [5, 86].

1.3.2 Angular Straggling. Radiation Length

Angular spread of the ion beam in passage through media is caused by *elastic* scattering of ions on medium particles when a well collimated beam becomes broadened on angles as shown in Fig. 1.1, middle. The angular straggling is also described by the Gaussian function.

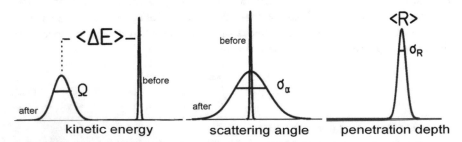

Fig. 1.1 Schematic changing of the incident beam energy and angular profiles due to interaction with media particles before and after penetration, and penetration depth. Left: energy loss $\langle\Delta E\rangle$ and stragling of the beam energy distribution with the width Ω. Middle: angular straggling—broadening over scattering angles with the distribution width σ_α. Right: penetration depth R (range) with the depth σ_R. From [81]

Multiple scattering of ions on media particles is considered in many theoretical [87–90] and experimental (e.g., [45]) works. In practice, for determination of the width σ_α of the Gaussian function, the following formula is used [88]:

$$\sigma_\alpha^2 = \frac{Z_1^2 \cdot 199 \cdot [\text{MeV}^2]}{(p\beta c)^2} \frac{L}{L_{rad}} \left[1 + 1/9 \cdot \log(L/L_{rad})\right] \ [\text{rad}^2], \qquad (1.5)$$

where L denotes the target thickness, Z_1 and p the charge and momentum of the incident ion, $\beta = \upsilon/c$, c the speed of light, and L_{rad} the *radiation length*.

The radiation length is a distance where the intensity of γ-radiation or flux of high-energy electrons is decreased on e times. The radiation length is given in g/cm^2, i.e., in the form independent on aggregation state of matter (liquid, gas, solid state) [89]:

$$L_{rad} \approx \frac{1433 \cdot M}{Z(Z+1)(11.32 - \ln Z)} \ [\text{g/cm}^2], \qquad (1.6)$$

where Z and M denote atomic number and atomic mass of the target atom. To get radiation length in units of cm, L_{Rad} value should be divided on material density. For example, for lead atoms one has: $Z = 82$, $M = 207$, $L_{rad} = 6.37$ g/cm^2, and L_{rad} (solid Pb, density 11.34 g/cm^3) = 0.57 cm.

1.3.3 Range (Penetration Depth)

In passage through matter, fast heavy ions lose their kinetic energy mainly due to ionization of the target particles and deviate slightly from their trajectory, which is nearly rectilinear. Therefore, the range (*penetration depth*) of heavy charged particle is defined by a distance from the impinging point to the point of the slow down to a stop. The range of the ion beam as well as other characteristics (energy and angular straggling) depend on the stopping power $-dE/dx$ and is defined by expression:

$$R = \int_0^{E_0} \left(-\frac{dE}{dx}\right)^{-1} dE, \qquad (1.7)$$

where E_0 denotes the initial beam energy. The quantity R is given in cm or g/cm^2 depending on units used for the stopping power.

Since the stopping power depends on the characteristics of the projectile ions and target atoms roughly as (see Sect. 1.2):

$$\left|\frac{dE}{dx}\right| \sim \frac{Z_1^2}{\upsilon^2} \rho, \qquad (1.8)$$

the penetration depth has the following properties

$$R \sim \frac{M v^4}{Z_1^2 \rho} \sim \frac{E^2}{M Z_1^2 \rho}, \tag{1.9}$$

where M, v, E, Z_1 denote mass, velocity, energy and charge of the projectile, respectively, and ρ the material density. In other words, at fixed velocity of the projectiles their range is proportional to their mass and inversely proportional to their squared charge, and at fixed energy are inversely proportional to their mass. Since with Z_1 and ρ increasing, the stopping power increases as in (1.8), the range in matter decreases. For example, for α-particles with energy of 10 MeV/u in air and aluminum foil, the ranges are 11.0 and 0.007 cm, respectively.

Information about ranges of ions in different media is of interest for many applications in radiation physics, biology, medicine etc. The data on ranges R can be found in [5, 12, 14, 91, 92].

Slowing-down time of a charged particle with kinetic energy E_0 is defined as

$$T_{stop} = \int_0^{E_0} \left(-v \frac{dE}{dx}\right)^{-1} dE. \tag{1.10}$$

1.4   Atomic Charge-Changing Processes in Gas/Solid Targets

Charge-changing processes arising in collisions of heavy ions with atoms (and molecules), are described by a general reaction called *transfer ionization* which involves simultaneous capture and ionization of both projectile and target electrons:

$$X^{q+} + A \rightarrow X^{q'+} + A^{m+} + (q' - q + m)e^-, \tag{1.11}$$

where q and q' denote charges of the projectile X before and after collision, respectively, and m the charge of the target A after collision.

Experimentally and theoretically the following elementary processes are investigated in more detail:

1. multiple-electron ionization of the projectile called *loss*, or *stripping*:

$$X^{q+} + A \rightarrow X^{(q+m)+} + \sum A + me^-, \quad m \geq 1, \tag{1.12}$$

where $\sum A$ means that the target can be excited or ionized,

2. target multiple-electron ionization called *target ionization*:

$$X^{q+} + A \rightarrow X^{q+} + A^{m+} + me^-, \quad m \geq 1, \tag{1.13}$$

3. multiple-electron capture also called *charge exchange*, or *electron transfer*, or *charge transfer*:

$$X^{q+} + A \rightarrow X^{(q-k)+} + A^{k+}, \ k \geq 1. \tag{1.14}$$

At high but non-relativistic collision energies $E < 200$ MeV/u, all these processes are accompanied by multiple-electron transitions that confirmed experimentally and theoretically. Besides pure theoretical interest to the nature of multiple-electron processes, they strongly contribute to the total (summed over m and k) cross sections and, therefore, should be taken into account together with single-electron processes, especially when heavy projectile and/or target atoms are involved. As the energy E increases to the relativistic region, a contribution of atomic multiple-electron processes decreases, and single-electron processes begin to play the main role.

At relativistic collision energies $E > 200$ MeV/u, besides a non-radiative capture (NRC) (1.14), one-electron *radiative electron capture* (REC), accompanied by a photon radiation, becomes important:

$$X^{q+} + A \rightarrow X^{(q-1)+} + A^+ + \hbar \omega_{REC}. \tag{1.15}$$

Radiative electron capture is similar to *radiative recombination* (RR), or *photo-recombination* process with one important difference: in REC the target *bound* electrons are captured meanwhile RR is a capture of *free* electrons, which usually takes places in plasmas.

At relativistic energies, the total capture cross section is given by the sum of both NRC and REC cross sections:

$$\sigma_{tot} = \sigma_{NRC} + \sigma_{REC}, \tag{1.16}$$

where a contribution of each process may be of the same order of magnitude, especially, in collisions of heavy ions with many-electron targets like Ar, Kr, Xe etc. (see [40, 42] and [43] for more detail).

At high non-relativistic energies E, single-electron cross sections of NRC, REC and electron loss (EL) reactions have the following asymptotic behavior:

$$\sigma_{NRC} \sim q^5 I_T^{5/2}/E^k, \ \ 1 \leq k \leq 5.5, \ \ v^2 >> I_T, \tag{1.17}$$

$$\sigma_{REC} \sim q^5 N_T/E^k, \ \ 1 \leq k \leq 2, \ \ v^2 >> I_T, \tag{1.18}$$

$$\sigma_{EL} \sim Z_T^2 \ln E/(q^2 E), \ \ v^2 >> I_P, \tag{1.19}$$

where I_P and I_T are the binding energies of the projectile ion and target atom, respectively, v is the projectile velocity, and Z_T and N_T are the nuclear charge and number of electrons of the target atom. The index k depends on the collision energy and atomic structure of electronic shells of colliding particles.

At present, there are a lot of data on experimental and theoretical electron capture and loss cross sections of reactions (1.12)–(1.14). At high energies, the experimental charge-changing cross sections of *uranium* ions (the heaviest natural element on Earth), colliding with various targets have been measured in a few laboratories: Super-HILAC at Lawrence Livermore National Laboratory, Princeton tokamak,

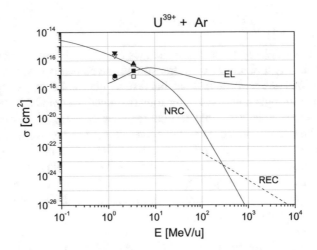

Fig. 1.2 EL, NRC and REC cross sections in collisions of U^{39+} ions with Ar atoms as a function of ion energy. Experiment: open symbols—one-electron processes, solid symbols—the total EL and NRC cross sections from [95] and [103]. Theory: solid curves—results by DEPOSIT and RICODE codes for EL and by CAPTURE code for NRC cross sections (see [42] for details); dashed curve denotes the REC cross sections calculated by the Kramers formula [114]. From [113]

A&M cyclotron at Texas, SIS/ESR at GSI, Darmstadt, and Radioactive-Isotope Beam Factory (RIBF) at RIKEN, Saitama, Japan. The corresponding experimental and theoretical data can be found in [93–113].

It is well known that at low collision energies electron capture is the dominant process, whereas electron loss is a minor one, particularly, for highly charged ions where most of electrons are tightly bound and hard to be ionized. With increasing ion energy E, electron-capture (EC) cross sections decrease rapidly ($\sim E^{-k}$, $k > 3$) due to the *velocity mismatch*, and then EL processes begin to play a major role in charge-changing collisions. It should be also noted that, above 500 MeV/u energy range, where practically no experiments have been reported so far, the EL cross sections show very weak dependence on the collision energy and finally tend to be more or less constant due to the relativistic effects. For molecular targets, the Bragg's additive rule is assumed so that the cross section for a molecule is represented as a sum of those for atoms composing the target molecule.

A typical example of pure electron-loss and capture cross-section behavior in collisions of U^{39+} ions with Ar atoms are shown in Fig. 1.2. At energies 100 keV/u $< E < 3$ MeV/u, NRC process plays the main role, and EL process dominates at $E > 20$ MeV/u. REC process becomes the main capture process at $E > 300$ MeV/u, i.e., at relativistic energies.

Chapter 2
Stopping Power of Ions in Matter (SP)

Abstract This chapter is devoted to properties of the *stopping power* (SP)—the important quantity characterizing the energy losses of projectile ions slowing down in gaseous, solid and plasma targets. The energy losses are resulted due to atomic processes between projectiles and target particles, mainly electron-loss, capture and target ionization. Energy and target-thickness (Bragg peak) dependencies of the SP are considered. Special attention is paid to the SP in plasmas as well as to the influence of the target-density effect on the SP values.

2.1 Introductory Remarks. Average Excitation Energy

In passing through matter, accelerated ions undergo thousands of collisions with the medium particles and loose their kinetic energy. Kinetic energy losses are characterized by the quantity $-dE/dx$ called *stopping power* of ions in matter [5]:

$$-\frac{dE}{dx} = \lim_{\Delta x \to 0} \frac{E_0 - E_1}{\Delta x} > 0, \qquad (2.1)$$

where $E_{0,1}$ denote ion kinetic energies before and after passing a target layer of the thickness Δx. The quantity (2.1) called a *linear* stopping power and has a dimension of eV/cm.

Energy losses in matter are also called *ionization losses* because in many cases they are caused by target-electron ionization. Information about ion stopping power (SP) are required for solving many problems in accelerator physics, controlled thermonuclear fusion, medicine etc. At present, SPs are investigated quite well experimentally for gaseous and solid targets (foils) in the energy range $E \approx 1\,\mathrm{keV/u}$ $-200\,\mathrm{GeV/u}$ for projectiles from protons to uranium (see, e.g., [92, 115–120]). At relativistic energies, the SPs are measured mainly at accelerators in BEVALAC, Berkeley [3], UNILAC, GSI, Darmstadt [62], and CERN [118].

© Springer International Publishing AG 2018
I. Tolstikhina et al., *Basic Atomic Interactions of Accelerated Heavy Ions in Matter*, Springer Series on Atomic, Optical, and Plasma Physics 98,
https://doi.org/10.1007/978-3-319-74992-1_2

Fig. 2.1 Recommended average relative excitation energies I/Z for neutral atoms with atomic numbers Z. From [119]

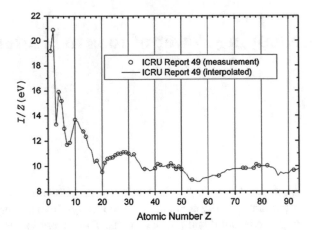

In the relativistic Born approximation SP of a heavy particle with the charge Z_1 and kinetic energy E, passing through gas target, has the form described by the Bethe-Bloch formula [16, 121]:

$$-\frac{dE}{dx} = \frac{4\pi e^4}{m} \cdot \frac{Z_1^2}{c^2\beta^2} \cdot N_e \cdot \left[\ln\frac{2mc^2\beta^2}{I} + \ln\gamma^2 - \beta^2\right], \qquad (2.2)$$

where m and e denote the electron mass and charge, $\beta = v/c$, N_e electron density in matter, γ the relativistic factor, I the *average excitation energy* of the target atom. The formula (2.2) was obtained taking into account the deviation of the precise theory from first-order quantum perturbation in terms of the momentum-transfer (transport) cross section for scattering of a free electron by the structureless ion with the charge Z_1.

Recommended average excitation energies I of neutral atoms are given in Fig. 2.1 as a function of the atomic number Z. In practice, the semiempirical Bloch formula [122] is used:

$$I \approx 10Z \text{ [eV]}, \qquad (2.3)$$

which gives the mean value of the data in Fig. 2.1.

Electron and atomic densities of matter, N_e and N_{at}, respectively, are related by

$$N_e = ZN_{at} = ZN_A\rho/M, \qquad (2.4)$$

where $N_A = 6.022 \times 10^{23}$ is the Avogadro number. Z, M and ρ denote atomic number, atomic mass (in a.m.u) and material density of matter (in g/cm^3). Atomic masses M are given in the Mendeleev periodic table of the chemical elements.

At non-relativistic energies, the SP in (2.2) reduces to the form:

$$-\frac{dE}{dx} = \frac{4\pi e^4}{mv^2} \cdot Z_1^2 \cdot Z \cdot N_{at} \cdot \ln\frac{2mv^2}{I}. \qquad (2.5)$$

This formula coincides with the one obtained in [17] on the basis of "effective slowdown" of ions in matter:

$$-\frac{dE}{dx} = N_{at}\kappa, \quad d\kappa = \sum_n (E_n - E_0)\,d\sigma_n, \tag{2.6}$$

where $E_{0,n}$ and σ_n denote energy levels of the ground and excited n-states and the excitation (ionization) cross sections to the n-states of the target atom, respectively. The sum is made over the discrete and continuous states. In derivation of (2.6), the first-order perturbation theory in dipole approximation is applied to the cross sections as well as the *sum rule* for the dipole oscillator strengths f_{on} [16]. Then the average excitation energy I is written in the form:

$$\ln I = \frac{\sum_n \ln f_{on}(E_n - E_0)}{\sum_n f_{on}} = \frac{1}{Z}\sum_n \ln f_{on}(E_n - E_0), \quad \sum_n \ln f_{on} = Z. \tag{2.7}$$

Along with the linear SP in (2.1), the *mass* stopping power

$$-\frac{1}{\rho}\frac{dE}{dx} \; [\text{MeV} \cdot \text{cm}^2/\text{g}], \tag{2.8}$$

is often used as it is independent on the material density ρ.

Experimental data show (see, e.g., [45]) that the (2.2) gives quite good results in cases when the collision parameter Z_1/υ is less than the unity:

$$\frac{Z_1}{\upsilon} = \frac{Z_1\alpha}{\beta} \leq 1, \tag{2.9}$$

where Z_1 and υ denote the charge and velocity of the projectile ion, $\alpha = 1/137$ is the fine-structure constant.

As the parameter Z_1/υ increases, the agreement between experimental SP data and the (2.2) becomes poorer. For a better description of experiment, the Lindhard-Sørensen (LS) approximation is used [121] based on the Dirac equation with account for the higher-order corrections such as the Bloch correction for close relativistic collisions [122], Barkas correction for polarization effects in media [123], Fermi correction for the density effects [124], the nuclear finite-size correction [121], shell-effect correction [119] and others.

Figure 2.2, left, shows experimental SPs for relativistic *bare* ions from oxygen to uranium in Be target as a function of the collision parameter, (2.9), at $\beta = 0.84$ in comparison with the Bethe-Bloch and the Lindhard-Sørensen models. With parameter Z_1/υ increasing, the Lindhard-Sørensen model gives a better agreement with experiment.

In the case of *dressed* heavy projectiles, the situation with SP behavior is not so clear: as the ion energy decreases, the projectile "effective" charge Z_{eff} becomes much smaller than its nuclear charge Z_N due to screening effects by projectile electrons and

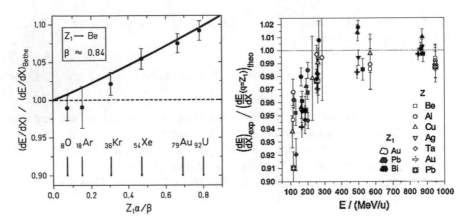

Fig. 2.2 Left: Relative SPs for bare nuclei from O ($Z_1 = 8$) to U ($Z_1 = 92$) projectiles in Be targets at $\beta = 0.84$ $\upsilon = 115$ a.u. as a function of the collision parameter (2.9). Circles—experiment, dashed line—Bether-Bloch formula (2.2), solid line—Lindhard-Sørensen approximation [121]. From [125]. Right: SP ratios for Au, Pb and Bi projectiles in solid targets from Be to Pb relative to those of the Lindhard-Sørensen approximation as a function ion energy. The charges of the projectile ions are close to their equilibrium values. From [126]

influence of atomic charge-changing processes—electron loss and electron capture. The existence of the screening effect is illustrated in Fig. 2.2, right, where the ratios of the experimental SPs to those calculated by the Lindhard-Sørensen model are shown for heavy Al, Pb and Bi ions in targets from Be to U as a function of ion energy. At energies $E < 300$ MeV/u, the ratios become smaller than unity indicating that $Z_{eff} < Z_N$ and that the electron screening becomes quite strong.

2.2   General Dependence of Stopping Power on the Ion Energy

On the basis of available experimental and theoretical data, the stopping power of ions in matter is presented as a sum of three components: interactions with target nuclei, electrons and photons. Schematically those dependencies on the ion energy is shown in Fig. 2.3 indicating the main atomic processes responsible for the SP in media.

At projectile energies $E < 1$ keV/u, the SP is caused by elastic collisions of projectile with the target *nuclei*. At $E > 10$ keV/u, slowdown of ions is due to atomic processes involving both projectile and target electrons. At ion velocities 1 a.u. $< \upsilon < \upsilon_e$ ($E > 25$ keV/u), where υ_e is the average electron velocity of the target atom, SP increases as $\sim\upsilon$ due to inelastic processes (electron capture and electron loss) in accordance with the Lindhard-Sørensen theory, reaches its maximum at $\upsilon \sim \upsilon_e$ and then falls down as $\sim \upsilon^{-2}$ until minimum value, given by the Bethe-Bloch formula

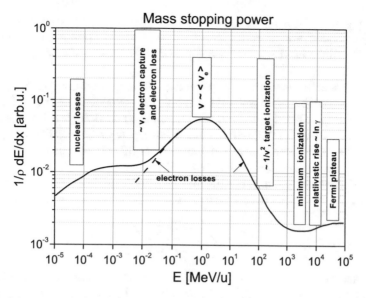

Fig. 2.3 Schematic behavior of the mass stopping power as a function of the projectile energy showing contributions of atomic processes at different energies

(2.2) is reached:

$$\left(-\frac{1}{\rho}\frac{dE}{dx}\right)_{min} \approx 1.7 \; \text{MeV} \cdot \text{cm}^2/\text{g}, \; \text{at} \; \beta\gamma = p/Mc \approx 3.5, \qquad (2.10)$$

where p and M denote momentum and mass of the projectile.

At energies $E \approx 10$–10^3 MeV/u the energy losses are associated with ionization of the target electrons, and the $\sim v^{-2}$ law corresponds to the asymptotic behavior of the ionization cross sections. At $E \geq 5$ GeV/u, the SP increases logarithmically $\sim \ln\gamma$ with energy showing the so-called *relativistic rise* described by the second log term in (2.2).

In the range of super high energies $E > 10^6$ MeV/u ($\beta\gamma \geq 100$), the relativistic rise of SP is cancelled due to the density correction and SP reaches the *Fermi plateau* [124]—a range of energies where SP no longer increases with increasing ion energy and stays nearly constant.

To calculate SP values, various computer codes used online such as SRIM (Stopping and Range of Ions in Matter) and TRIM (TRansport of Ions in Matter) [127], MSTAR [128, 129] and others.

2.3 Projectile Effective Charge

One of the relevant questions arising in passage of projectile ions in matter is the problem of the charge state of ions inside matter. Since this value can not be measured, two values are introduced: the *equilibrium* $\bar{q}$ and the *effective* Z_{eff} charges of penetrating ions. The $\bar{q}$ values are determined on the basis of competition of ionization and recombination processes, or electron-loss and electron-capture in gas/solid targets. The difference between $\bar{q}$ and Z_{eff} charges is discussed in Sect. 3.6 (see also [130–137]).

Experimental SP data for various projectiles and targets are given in [5, 138–146], and also are available in websites of the NIST [92]—National Institute of Standards and Technology, ICRU [119]—Intl. Commission on Radiological Units and Measurements, and IAEA [120]—Intl. Atomic Energy Agency. Detailed comparative analysis of available experimental and theoretical SP data of ions from Li to Kr at energies 0.001–1000 MeV/u is given in [145].

The effective charge of projectile ions is found experimentally from measured SPs of ions in question and reference ions, usually protons or α-particles, in the form:

$$Z_{eff}^2 = Z_0^2 \, \frac{dE/dx(\upsilon, Z_N)}{dE/dx(\upsilon, Z_0)}, \tag{2.11}$$

where υ and Z_N denote the projectile velocity and nuclear charge, and Z_0 the effective charge of the reference ions: $Z_0 = 1$ for protons and $Z_0 = 2$ for α-particles. Equation (2.11) is based on the perturbation theory, according to which the SP of bare ions with the charge Z_N is proportional to Z_N^2 if the ion velocity satisfies the *Bohr criterion* (Sect. 3.4):

$$Z_N e^2 / \hbar \upsilon \ll 1. \tag{2.12}$$

The condition (2.12) is not fulfilled in many cases, in which the quantity Z_{eff} does not play a role, e.g., in slow collisions where the SPs undergo strong oscillations (see [146]). In general, calculations of the ion effective charge require non-perturbative quantum calculations including higher-order terms (see, e.g., [147–149]).

In many cases, the use of semiempirical formulae of the projectile effective charge allows one to describe and predict the SPs quite effectively for heavy ions in a wide energy range. The following formula, which has the exponential form (the Thomas-Fermi scaling) suggested in [5, 140] using protons as the reference ion is often used for Z_{eff}:

$$Z_{eff}(\upsilon) = Z_N \left[1 - \exp\left(-0.92 Z_N^{-2/3}\right) \right], \tag{2.13}$$

where υ and Z_N are the velocity and charge of the projectile. According to compilations [5], (2.13) reproduces experimental data within 10% at ion velocities $\upsilon > 3$ a.u.

As for α-particle, a fitting formula suggested in [5], is used at any projectile energy:

$$Z_{eff}/2 = 1 - \exp\left(-\sum_{i=0}^{5} a_i \, \ln^i(E)\right), \tag{2.14}$$

where the ion energy E is in keV/u and the fitting coefficients a_0 through a_5 are equal to 0.2865, 0.1266, 0.001429, 0.02402, 0.01135, and 0.00175, respectively.

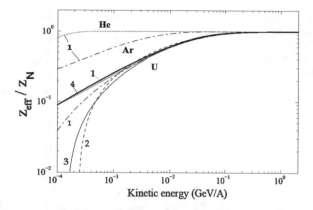

Fig. 2.4 Calculated effective charges Z_{eff} of light and heavy projectiles relative to their nuclear charge Z_N in aluminum foil. Numbers near curves indicate the references used for the calculation of Z_{eff}: 1—[5], 2—[138], 3—[132] and 4—[130]. From [136]

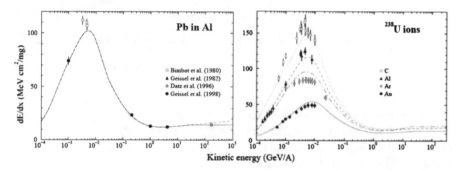

Fig. 2.5 Mass stopping power in Ar gas and foils as a function of ion energy. Symbols: experiment [62, 150–152]; curves: theory—the mix-and-match procedure [136] using a point-like projectile. Left: mass SPs for lead ions in aluminum. Right: mass SPs for uranium ions in solid (C, Al and Au) targets. From [136]

Calculated ratios of ion effective charge to the nuclear charge of He, Al and U projectiles in collisions with Al foil are presented in Fig. 2.4 as a function of the ion energy and nuclear charge using different approximation formulae. As seen from the figure, the effective charge increases with the ion nuclear charge.

Experimental data on SPs for ions with the nuclear charge $2 \leq Z_N \leq 103$ and energies $E = 0.0125$–12 MeV/u in gases and foils are given in [89]. There, SPs for heavy ions are obtained using semiempirical formula (2.11) with protons as reference ions. Since experimental data at that time were quite limited, the accuracy of the data in [89] in some cases is not better than about 50%.

Mass SPs of heavy ions in foils and Ar gas, calculated by the MARS Monte Carlo code [136], are presented in Fig. 2.5, in comparison with available experimental data. Calculations include various corrections to dE/dx at low and high energies.

Fig. 2.6 Penetration depth profiles of incident flux of photons of different wavelengths and carbon ions (Bragg peaks for 250 and 300 MeV/u) in passage through water. From [7]

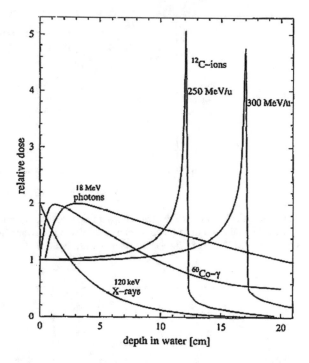

In calculations the *mix-and-match* procedure is employed providing a good, within 10%, agreement with experiment at low energies. The experimental data for Al foil at 160 GeV/u [152] in Fig. 2.5, right, corresponds to the highest energy achieved for accelerated heavy ions at present time.

2.4 Bragg Peak

One of the important properties of the SP is its dependence on the target *thickness* manifesting in the *Bragg peak* [153]. In passage of fast heavy ions through a thick target, they lose their energy and the SP drastically increases due to increase of the target ionization cross section proportionally to the square of the ion velocity $\sim v^2$, which results in a sharp peak at a certain target thickness, called a particle *range* (Sect. 1.7). As an example, the penetration depth profiles for short-wavelength photons and carbon ions in a water target are shown in Fig. 2.6.

The presence of the Bragg peak on the ionization loss curves is widely used in practice, mainly in cancer therapy (see [7, 8, 13]) using proton beams or heavier ions (C^{q+}, N^{q+}) as well as photons. In the case of beam therapy, the radiation dose peak gives rise to deeper penetration depths compared to photon irradiations; moreover, intensity and the position of the Bragg peak can be controlled by ion species and energy. The latter gives large advantages of using ion therapy compared to photon radiation with short and ultra short wavelengths as is demonstrated in Fig. 2.6.

2.5 SP of Heavy Ions in Plasmas

Theoretical problems of slowdown of fast heavy ion beams in a cold plasma are considered in many works (e.g., [74, 75, 137, 154–161]. The projectile stopping power on plasma free electrons has the form [154] [(cf. (2.5)]:

$$-\left[\frac{dE}{dx}\right]_{free} = \left(\frac{Z_{eff}\, e\, \omega_p}{\upsilon}\right)^2 \cdot \ln\left(\frac{m\upsilon^3}{Z_{eff}\, e\, \omega_p}\right), \quad \upsilon \gg \upsilon_{th} = 1.13\sqrt{kT_e/m}, \quad (2.15)$$

where T_e and N_e denote the electron temperature and density, respectively, k the Boltzman constant, υ the projectile velocity.

The plasma frequency ω_p is given by:

$$\omega_p = \left(4\pi N_e e^2/m\right)^{1/2}. \quad (2.16)$$

For non-relativistic ions penetrating gaseous targets, (2.5) can be rewritten in the form similar to eq. (2.15):

$$-\left[\frac{dE}{dx}\right]_{gas} = \left(\frac{Z_{eff}\, e\, \omega_p}{\upsilon}\right)^2 \cdot \ln\left(\frac{2m\upsilon^2}{I}\right), \quad (2.17)$$

where I is the average excitation energy (Sect. 2.1) and N_e in (2.16) is the density of the *bound* electrons belonging to plasma atoms, ions and molecules.

Equations (2.15) and (2.17) differ from each other by the logarithmic function, that means that even in the case of equal effective charges Z_{eff} in a cold gas and a fully ionized plasma, SP on free electrons in a plasma is always greater than that in a gas target. For *partly* ionized plasma, ion energy losses are defined by a sum of the terms (2.17) for bound electrons and (2.15) for free electrons, and each term should be multiplied on the corresponding density of particles in the plasma.

The presence of free electrons significantly changes the character of atomic interactions of ion beams with plasmas compared with a cold gas target. Along with free electrons in a plasma, there is a fraction of positive ions, and, in addition, densities of each particle components depend on the plasma temperature and density. Besides electron loss and capture processes involving neutral atoms in gas/solid targets, interaction with plasma free electrons leads to additional atomic processes such as radiative recombination (capture of free electron accompanied by photon emission), dielectronic recombination, three-body recombination, ionization of ions by electron and ion impact and so on (see Chap. 7).

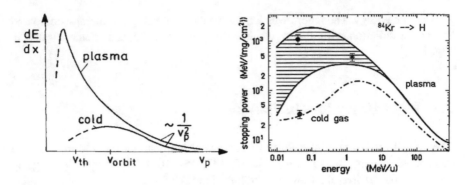

Fig. 2.7 Left: Schematic illustration of increasing the ion SP in plasmas compared to cold gas as a function of ion velocity provided the ion effective charge is the same for both cases. From [74]. Right: SPs of Kr-ions in a fully ionized hydrogen plasma and a cold gas as a function of the ion energy. Experiment—symbols, theory—solid and dash-dotted curves. From [76]

Reaction rates of recombination processes on free electrons are much smaller than those of electron capture processes on bound electrons which leads to significant increase of ion effective charge Z_{eff} in plasma targets compared to gas targets of the same element. This increase, predicted in [155], also causes an increase of the ion stopping power because in the first approximation, the stopping power is proportional to Z_{eff}^2.

Experimental investigations of SPs of heavy ions were carried out in [162–170]. For example, in [162, 170] a hydrogen plasma was used in a Z-pinch machine with electron density $N_e \sim 10^{16} - 10^{19}$ cm^{-3} and temperature $T_e \approx 10$–20 eV.

An increase of ion SP in a fully ionized plasma compared to a cold gas is illustrated in Fig. 2.7, left. As has been mentioned before, for a fully ionized plasma the SP is greater because of the different logarithmic dependencies: free electrons in a plasma are much easier to be excited (plasma waves) than the bound electrons in atoms and ions; this conclusion is confirmed by experimental data in [162, 170, 171].

Figure 2.7, right shows experimental data on SPs for Kr ions in a fully ionized hydrogen plasma and a cold gas as a function of the ion energy. At rather low energy $E \sim 0.1$ MeV/u SP in a hydrogen plasma is about 200 times larger than in a hydrogen cold gas whereas at high energies $E > 10$ MeV/u, the difference in SPs for a plasma and a gas is a factor of 2.

At present, intensive investigations are being carried out on slowdown of heavy ions in a *laser-produced* carbon plasmas with higher electron density $N_e \sim 10^{21}$ cm^{-3} and temperature $T_e = 60$–250 eV (see [163–188]). Higher N_e and T_e plasma parameters and a presence of complex ions with different charges (C^{q+}) allow one to perform more detailed investigations of atomic processes of ion beams with plasmas such as dielectronic recombination that can not occur in a hydrogen plasma with creation of doubly-excited states [(see (7.2)].

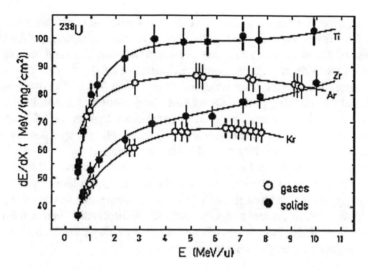

Fig. 2.8 Experimental stopping powers of uranium ions in solid targets (Ti, Ar) and nearby gases (Kr, Ar) as a function of ion energy. The lines are fits to the data. From [115]

2.6 Influence of the Target-Density Effect on SP in Plasmas

The density effect, or *target-density* or *gas-solid* effect, was discovered experimentally [175] in investigation of the charge-state fractions of uranium ion beams passing through carbon foils and gaseous targets, and later—in measurements of the ion stopping powers in gaseous and solid media [115, 176]. The density effect consists in the increase of the *equilibrium mean* charge (Sect. 3.6) of the ion beam penetrating through solid targets compared to the gaseous targets.

First theoretical models of the target-density effect on *charge-state fractions* and stopping powers were presented in the works [177–181]. With development of powerful accelerators of heavy ions, experimental and theoretical investigations of the density effect were continued (see [1, 5, 151–185]). At present, the term *density effect* is treated more widely, i.e., as an influence of the target density on the charge-changing cross sections, on the stopping power in a dense media, on equilibrium mean charges in passage of ion beam through gas, solid and plasma targets etc.

Qualitatively the effect can be explained in the following way (see also Sects. 4.3 and 6.6). As the target density increases, the collision frequency of accelerated ions with media particles also increases so that the time between successive collisions becomes shorter than the lifetimes of the ion excited states and a part of ions, being in the excited states, undergoes further collisions with the target particles. Excited ions have no time to make transitions into lower and the ground states by radiative decay or some other way that leads to ionization of these ions by subsequent collisions with the target particles.

As a result, with increasing the target density, electron-capture cross sections decrease because the number of the vacant excited states decreases, but electron-loss cross sections, on the contrary, increase due to ionization not only from the ground but also from closest excited states. Because the average mean charge of the incident ion is created due to a balance between ionization (electron loss) and recombination (electron capture) processes, the combined influence of the density effect on the both processes leads to the increase of the mean charge in more dense media and, therefore, to the stopping power. Quantitative explanation of the influence of the density effect on the ion mean charge in a dense media is given in [182] in terms of the electron-loss and capture cross sections, depending on the target density, relative collision energy and atomic structure of colliding particles.

The first measurements of SPs of heavy many-electron ions, confirming the influence of the density effect, were carried out at the UNILAC accelerator at GSI, Darmstadt, using partly ionized ions from Kr to U energies around a few MeV/u [151]. It was shown that SPs in gaseous targets are about 20–30% smaller than in solid targets which is shown in Fig. 2.8 for SPs of uranium ions passing through heavy gaseous and solid targets with neighboring atomic numbers.

Chapter 3
Evolution of the Projectile Charge-State Fractions in Matter

Abstract The key question arising in passage of ion beams through media is the evolution of the projectile *charge-state fractions* $F_q(x)$ in matter as a function of the target thickness x. Experimental and theoretical information on $F_q(x)$ values are required for solving many problems in atomic, plasma and accelerator physics. For example, electron-loss and capture cross sections are usually determined by measured equilibrium charge-state fractions [1], [186]. After a number of subsequent collisions with the target particles, the charge-state distribution becomes dynamically stable and reaches its *equilibrium* with an average (mean) charge state $\bar{q}$. This chapter is devoted to determination of equilibrium and non-equilibrium charge-state fractions on the basis of differential balance rate equations with coefficients equal to the interaction charge-changing cross sections. The Allison three-charge-state model as well as different computer codes are considered. To solve the balance rate equations, semi-classical and semiempirical formulae for the equilibrium mean charges $\bar{q}$ are given for ion beams passing through gaseous and solid targets.

3.1 Balance Rate Equations for Charge-State Fractions. Equilibrium Regime

The charge-state fraction $F_q(x)$ is characterized by a probability of the projectile ion to be in a certain charge state q at a target thickness x. As the target thickness x increases, $F_q(x)$ values strongly change due to the influence of competing ionization (electron loss) and recombination (electron capture) processes, i.e., due to charge-changing processes occurring in media. As a rule, x-dependence of the fractions $F_q(x)$ is found by solving first-order partial differential *balance* rate equations in which the coefficients are equal to charge-changing cross sections.

© Springer International Publishing AG 2018
I. Tolstikhina et al., *Basic Atomic Interactions of Accelerated Heavy Ions in Matter*, Springer Series on Atomic, Optical, and Plasma Physics 98,
https://doi.org/10.1007/978-3-319-74992-1_3

For *gas/foil* targets, the rate equations are written in the form [1]:

$$\frac{dF_q(x)}{dx} = \sum_{q' \neq q} F_{q'}(x) \cdot \sigma_{q'q} - F_q(x) \sum_{q' \neq q} \sigma_{qq'}, \tag{3.1}$$

$$\sum_q F_q(x) = 1, \quad x = NL, \tag{3.2}$$

where x denote the target thickness or an *areal density* having a dimension of atom/cm^2 or molecule/cm^2, N the target density, L the penetration depth or *effective depth* in cm, σ_{ij} multiple-electron loss cross sections (projectile ionization by target particles) at $i < j$, and multiple electron capture cross sections at $i > j$. The sum on q' is made over cross sections with all possible charge states q.

It is assumed that the charge-changing cross sections σ_{ij} are independent on the projectile energy within considered target thickness x. This dependence arises in a dense matter (dense gas or foil), e.g., when the influence of the *target-density effect* should be accounted for (see Sects. 4.3 and 6.6).

The set of equations (3.1) and (3.2) is solved numerically using the Runge-Kutta method or matrix-diagonalization method (Sect. 3.7). In a *plasma* target, the equation system (3.1) and (3.2) is solved using the *rate coefficients* $N \upsilon \sigma$ instead of the cross sections (N denotes a density of different plasma particles, and υ the average velocity of ions) and with account for additional atomic processes induced by plasma free electrons and ions: radiative and dielectronic recombination, ionization by electron and ion impact and others which are absent in case of gas/solid targets (see, e.g., [75, 187, 188] and Chap. 7).

The areal density x can be used in different units through the relation:

$$x[\text{atom/cm}^2] = N[\text{atom/cm}^3] \cdot L[\text{cm}] = x[\text{g/cm}^2] \frac{N_A}{M}, \tag{3.3}$$

where $N_A = 6.022 \times 10^{23}$ is the Avogadro number and M is the target particle (atom or molecule) mass number in atomic units.

Charge-state fractions $F_q(x)$ in media have a significant property: at a certain target thickness, called *equilibrium thickness* x_{eq}, all fractions become nearly independent on the target thickness and reach their equilibrium stages, called *equilibrium* fractions $F_q(\infty)$. Therefore, at $x > x_{eq}$ one has:

$$F_q(x > x_{eq}) = \text{constant} \equiv F_q(\infty). \tag{3.4}$$

The equilibrium thickness depends on the ion energy and interaction cross sections of the ions beam with the target particles according to (3.1) and (3.2) and, generally speaking, on the charge of incoming ions [14].

Fig. 3.1 Evolution of charge-state fractions of bromine ions formed in collisions of Br^{10+} with Ar at energy 13.9 MeV (174 keV/u) as a function of Ar-atom thickness. Symbols—experimental data, curves—calculation with account for multiple-electron loss processes. Charge states q of bromine ions are indicated on the right-hand side of the figure. From [189]

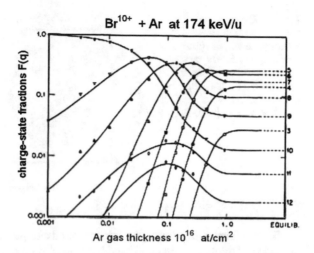

An example of evolution of the charge-state fraction is shown in Fig. 3.1 for 174 keV/u Br^{10+} ions in collisions with Ar gas as a function of the gas thickness. Experimental data are indicated by symbols and calculated fractions by solid curves. The initial conditions for $F_q(x)$ values are shown at $x = 0$, and the charge states q are indicated on the right side of the figure. At target densities (thickness) $x \geq x_{eq} \approx 3 \times 10^{16}$ atom/cm^2 all fractions are saturated independently on the further increasing of the thickness.

The *mean* charge of the ion beam at the target thickness x is defined as:

$$q(x) = \sum_q qF_q(x), \quad \sum_q F_q(x) = 1. \tag{3.5}$$

Equilibrium mean charge is given by

$$\bar{q} \equiv q(\infty) = \sum_q qF_q(\infty), \tag{3.6}$$

where $F_q(\infty)$ are the equilibrium fractions. The $\bar{q}$ value needs not to be an integer number.

The mean charge $q(x)$ keeps constant in the *equilibrium* regime, i.e., $q(x > x_{eq}) = \bar{q}$. This is a very important property of the ion beam—target interactions which is used for optimization of the mean charge of exit ions. Figure 3.2 shows experimental dependence of the mean-charge $q(x)$ on the carbon-foil thickness x for collisions of 2 MeV/u - S^{q+} ions with initial charges $q_0 = 6$–16 and ion energy of 2 MeV/u. Curves of the mean charges $\bar{q}(x)$ converge into one point corresponding to the equilibrium charge $\bar{q} \approx 12.7$ at $x_{eq} \approx 200 \, \mu g/cm^2$ even for ion beams with initial charge states $q_0 = 14 - 16$, i.e., higher than $\bar{q}$.

Fig. 3.2 Experimental
dependence of the mean
charge state $q(x)$ of sulfur
ions as a function the target
thickness x in collisions of
S^{q+}, $q = 6$–16, ions with
C-foil at energy of
2.0 MeV/u. From [190, 191]

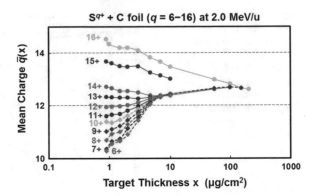

We note that the mean charge states for initial charge states $q_0 = 6 - 14$ once merge
at the target thickness $6.9 \mu/cm^2$ and increase simultaneously as the target thickness
is further increased, whereas those for $q_0 = 15$ and 16 evolve straight forwardly to the
equilibrium. This phenomenon is also described by the interaction cross sections of
ions with different charge states, which the authors of [190] call a *quasi-equilibrium*.

3.2 Allison Three-Charge-State Model

In the case of three-state system, where fractions are F_0, F_1 and F_2, the rate equations
(3.1) and (3.2) can be solved analytically as was shown by Allison in [186]. There,
the balance equations are presented in the form:

$$\frac{dF_0}{dx} = -F_0(\sigma_{01} + \sigma_{02} + \sigma_{20}) + F_1(\sigma_{10} - \sigma_{20}) + \sigma_{20}, \tag{3.7}$$

$$\frac{dF_1}{dx} = F_0(\sigma_{01} - \sigma_{21}) - F_1(\sigma_{10} + \sigma_{12} + \sigma_{21}) + \sigma_{21}, \tag{3.8}$$

$$F_2 = 1 - (F_0 + F_1), \tag{3.9}$$

i.e., when the system includes single- and double-electron loss cross sections σ_{01}, σ_{12}
and σ_{02}, and capture cross sections σ_{10}, σ_{21} and σ_{20}, respectively. Analytical solution
of (3.7)–(3.8) is presented in [186] in a close analytical form as a function of the
target thickness x, using initial condition $F_q(x = 0)$ and *equilibrium* charge-state
values $F_q(\infty)$ for $q = 0, 1, 2$.

In the case of two charge-state fractions F_1 and F_2, the solution of (3.7)–(3.8) has
the form:

$$F_i(x) = F_i(\infty) + (F_i(0) - F_i(\infty)) \cdot \exp\left[-(\sigma_{12} + \sigma_{21})x\right], \quad i = 1, 2, \tag{3.10}$$

$$F_1(\infty) = \frac{\sigma_{21}}{\sigma_{12} + \sigma_{21}}, \quad F_2(\infty) = \frac{\sigma_{12}}{\sigma_{12} + \sigma_{21}}, \quad F_1 + F_2 = 1. \tag{3.11}$$

For two cases with different initial conditions, one has:
(1) $F_1(0) = 0$, $F_2(0) = 1$:

$$F_1(x) = \frac{\sigma_{21}}{\sigma_{12} + \sigma_{21}} \left(1 - \exp\left[-(\sigma_{12} + \sigma_{21}) x\right]\right), \tag{3.12}$$

$$F_2(x) = \frac{\sigma_{12}}{\sigma_{12} + \sigma_{21}} + \frac{\sigma_{21}}{\sigma_{12} + \sigma_{21}} \exp\left[-(\sigma_{12} + \sigma_{21}) x\right]. \tag{3.13}$$

(2) $F_1(0) = 1$, $F_2(0) = 0$:

$$F_1(x) = \frac{\sigma_{21}}{\sigma_{12} + \sigma_{21}} + \frac{\sigma_{12}}{\sigma_{12} + \sigma_{21}} \exp\left[-(\sigma_{12} + \sigma_{21}) x\right], \tag{3.14}$$

$$F_2(x) = \frac{\sigma_{12}}{\sigma_{12} + \sigma_{21}} \left(1 - \exp\left[-(\sigma_{12} + \sigma_{21}) x\right]\right). \tag{3.15}$$

The cases (1) and (2) are considered in [184] for investigation of the *density effect* (Sect. 6.6), in which fractions of bare and H-like nickel ions colliding with solid and gaseous targets at 200 MeV/u were measured as a function of the target thickness (see Fig. 3.3). Experimental data are compared with theoretical two-component model for H-like and bare nickel ions, Ni^{27+} and Ni^{28+} using (3.12)–(3.15). Electron-loss and capture cross sections are calculated in relativistic approximation with account for the density effect. A gas-solid effect was clearly demonstrated experimentally and theoretically using the two-component model for non-equilibrium charge-state fractions.

Allison analytical three-component model is often used in case of fast ions with large charge states $q \gg 1$ when only three charge states are dominant, i.e., bare, H-

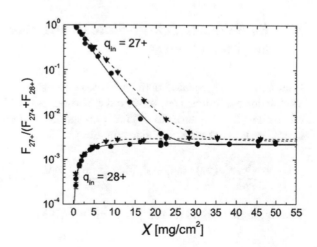

Fig. 3.3 Charge-state fractions for 200 MeV/u Ni^{27+} ions emerging from targets of solid (polypropylene $((C_3H_6)_n)$), triangles, and gaseous (ethylene C_2H_4), circles, targets as a function of the target thickness x for incoming beams of Ni^{27+}, $q_{in} = 27+$ and of Ni^{28+}, $q_{in} = 28+$. The solid and dashed curves represent the least-squares fits to (3.14) and (3.12). From [184]

Fig. 3.4 Evolution of
charge-state fractions of
950-MeV/u bare U ions
impinging on Ti targets.
Symbols—experiment,
curves—results of the
CHARGE code. From [117]

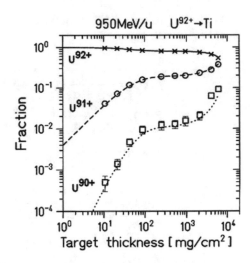

and He-like ions. The three-state model is also used when the projectile is stripped
up to the K-shell and the *Bohr criterion* is fulfilled (Sect. 3.4).

The analytical solution for three-state systems was used in [192] to calculate non-
equilibrium fractions and charges for collisions of light ions with celluloid at low
energy (0.75 MeV/u). Another example of using the three-component model is the
CHARGE code [117], which is intended to describe the charge-state evolution in
the high-energy domain where only bare, H-, and He-like ions present (Sect. 3.7).
The last example is shown in Fig. 3.4, where evolution of charge-state fractions of
950-MeV/u bare U ions impinging on Ti targets are presented as a function of the
target thickness.

3.3 Equilibrium Charge-State Distribution and Mean Charge

Equilibrium charge-state fractions $F_q(\infty)$ correspond to the equilibrium regime of
the fraction evolution. They are found experimentally or from the balance equations,
when all derivatives turns to $dF_q/dx = 0$ and the system (3.1) and (3.2) becomes a
simple system of linear algebraic equations:

$$\sum_{q' \neq q} F_{q'}(\infty) \cdot \sigma_{q'q} - F_q(\infty) \sum_{q' \neq q} \sigma_{qq'} = 0, \quad \sum_q F_q(\infty) = 1. \quad (3.16)$$

The equilibrium mean charge $\bar{q}$ is given by eq. (3.6).

The algebraic system (3.16) has a simple analytical solution if only single-electron processes are accounted for, i.e., $|q - q'| = 1$. Then the equilibrium fractions $F_q(\infty)$ are expressed as the ratios of single-electron loss-to-capture cross sections [1, 117]. For example, for a four-state model the solution of (3.16) has the form:

$$F_1(\infty) = \cfrac{1}{1 + \frac{\sigma_{12}}{\sigma_{21}}\left(1 + \frac{\sigma_{23}}{\sigma_{32}}\left(1 + \frac{\sigma_{34}}{\sigma_{43}}\right)\right)}, \tag{3.17}$$

$$F_2(\infty) = F_1(\infty)\,\frac{\sigma_{12}}{\sigma_{21}},$$

$$F_3(\infty) = F_2(\infty)\,\frac{\sigma_{23}}{\sigma_{32}},$$

$$F_4(\infty) = 1 - [F_1(\infty) + F_2(\infty) + F_3(\infty)],$$

where $\sigma_{12}, \sigma_{23}, \sigma_{34}$ denote single-electron loss cross sections for transitions $q \to q+1$, and $\sigma_{21}, \sigma_{32}, \sigma_{43}$ single-electron capture cross sections for transitions $q+1 \to q$. Analytical expressions for the equilibrium charge-state fractions can be easily generalized for the case of an arbitrary number of the charge states taken into account.

Experimental equilibrium charge-state distributions (CSD) of $F_q(\infty)$ values over q are presented in Fig. 3.5 for collisions of Cu ions with Mo foil at energy of around 1.8 MeV/u [193, 195]. As it is seen, two different measurements are in a very good agreement with each other.

As was mentioned above, the equilibrium fractions $F_q(\infty)$ depend on the interaction cross sections between projectile ions and target particles but are independent on the charge state q_0 of incident ions. This feature follows from experimental data (see, e.g., [190]) and theory [14], and is widely used in applications, e.g., for detection of super-heavy elements (Sect. 9.3).

Fig. 3.5 Equilibrium CSD of Cu ions in Mo measured in [195], inverted triangles and in [193], upright triangles. From [193]

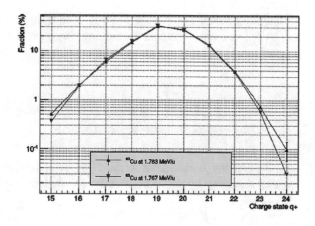

3.4 Bohr Semi-classical Velocity Criterion

There are various semi-classical and semiempirical formulae to estimate the equilibrium mean charge $\bar{q}$ of fast ion beams penetrating gas/solid target. We note that although these formulae are quite simple and very useful for many applications, they have some serious disadvantages, namely, they do not account for atomic structure of colliding particles and the target-density effect.

First semi-classical formulae for $\bar{q}$ values were obtained independently in the works of Bohr [179] and Lamb [194] (see also [14, 181] for more details). The Bohr criterion for equilibrium ion charge in matter was obtained for dilute gaseous targets assuming that the projectile loses all electrons whose, orbital velocity v_e exceeds the ion velocity, v: $v_e > v$. The Bohr formula was obtained using the semi-classical atom model and two expressions:

$$v_e \approx q/n, \quad n \approx Z_N^{1/3}, \tag{3.18}$$

where q and Z_N denote projectile ion and nuclear charges, respectively, and n the principal quantum number. With these approximations, the Bohr criterion is written in the form:

$$\bar{q}/Z_N \approx v/Z_N^{2/3}, \quad 1 < v < Z_N^{2/3}, \tag{3.19}$$

where $Z_N^{2/3} = v_{TF}$ is an electron velocity in the Thomas-Fermi model. According to (3.19), the total loss of projectile electrons, forming bare ions, occurs at ion velocity $v \approx Z_N^{2/3}$.

Since the ratio $\bar{q}/Z_N$ can not exceed the unity, the Bohr formula can be applied only for ion velocity given in (3.19). In order to extrapolate the Bohr formula to higher ion velocities, the formula (3.19) was modified in the form [130]:

$$\bar{q} \approx Z_N \left(1 - e^{-v/Z_N^{2/3}}\right). \tag{3.20}$$

The Bohr formula (3.19), although derived from simple physical assumptions, is still useful as a basic formula for solving many problems in the field of heavy-ion penetration.

3.5 Semi-empirical Formulae for Equilibrium Mean Charges. Gaussian Distribution

Equilibrium charge-state distribution (CSD) of fractions $F_q(\infty)$ over charge-state q are well-described by the Gaussian function with the width d:

$$F_q(\infty) = 1/(\sqrt{2\pi}\, d) \cdot \exp\left[-(q - \bar{q})^2/(2d^2)\right], \tag{3.21}$$

$$d = \left[\sum_q (q - \bar{q})^2 F_q(\infty) \right]^{1/2}, \tag{3.22}$$

and the asymmetry parameter s (skewness):

$$s = \sum_q \frac{(q - \bar{q})^3 F_q(\infty)}{d^3}. \tag{3.23}$$

Yielding the equilibrium mean charge $\bar{q}$ from the balance rate equations (3.1) and (3.2) is a quite complicated problem because of the necessity to know a lot of charge-changing cross sections and to use a limited number of equations. Therefore, in practice, the $\bar{q}$ values are found experimentally from the crossing point of two curves of ionization and recombination cross sections as a function of an ion charge q, i.e., from equality:

$$\sigma_{ion}(\bar{q}) = \sigma_{rec}(\bar{q}). \tag{3.24}$$

This method gives approximate estimates for $\bar{q}$ compared to (3.6) but is simpler and clearer. The difference in $\bar{q}$ values obtained by (3.24) and (3.6) can be of the order of 20–30% (see [196, 197]).

Below, various frequently used semiempirical formulae are given the most often used for the mean equilibrium charge in applications. A recent review and evaluation of the relevant semiempirical models for mean charge $\bar{q}$, distribution width d and skewness s are presented in [191, 193].

The Betz et al. formula [198] was obtained for *gaseous* and *solid* targets on the basis of experimental data for projectile ions with the nuclear charge $Z_N > 10$ and energies $E = 5-80\,\text{MeV}$:

$$\bar{q} = Z_N \left[1 - C e^{-\upsilon Z_N^{-\gamma}} \right], \quad \upsilon > 1, \tag{3.25}$$

where the fitting parameters $C \approx 1$ and $\gamma \approx 2/3$. At $\upsilon Z_N^{-2/3} \ll 1$, (3.25) coincides with the Bohr formula (3.19).

The Nicolaev and Dmitriev formula [199] was derived for heavy many-electron ions with energy $E > 100$ MeV penetrating *solid* targets (foils) and has the from:

$$\bar{q} = Z_N \left[1 + \left(\frac{0.608\upsilon}{Z_N^a} \right)^{-1/k} \right]^{-k}, \quad \upsilon > 1, \tag{3.26}$$

with fitting parameters $a = 0.45$ and $k = 0.6$.

For carbon-foil targets, the Shima et al. formula [200] is often used:

$$\bar{q} = Z_N \left(1 - e^{-1.25x+0.32x^2-0.11x^3} \right), \quad x = 0.608\upsilon Z_N^{-0.45}, \quad x < 2.4. \tag{3.27}$$

Another semiempirical formula suggested by Baron et al. [201] for ions with $18 \leq Z_N \leq 92$ and energies $E < 10.6\,\text{MeV/u}$ in carbon foils is also used:

$$\bar{q} = Z_N \left[1 - C \cdot \exp\left(-83.275\beta/Z_N^{0.477}\right)\right] \tag{3.28}$$
$$\times \left[1 - \exp\left(-12.905 + 0.2124Z_N - 0.00122Z_N^2\right)\right],$$
$$C = 0.9 + 0.0769E, \quad E < 1.3\,\text{MeV/u}, \tag{3.29}$$
$$C = 1, \quad 1.3\,\text{MeV/u} < E < 10.6\,\text{MeV/u}. \tag{3.30}$$

For gaseous and solid targets, the Schiwietz formulae are used [202]:

$$\bar{q}_{gas} = Z_N \frac{376x + x^6}{1428 - 1206x^{0.5} + 690x + x^6}, \quad x = \left(\frac{a}{Z^{0.017a - 0.03}}\right)^{1 + 0.4/Z_N}, \quad a = \upsilon Z_N^{-0.52}, \tag{3.31}$$

for *gaseous* targets and

$$\bar{q}_{solid} = Z_N \frac{12x + x^4}{0.07/x + 6 + 0.3x^{0.5} + 10.37x + x^4}, \quad x = \left(\frac{a}{Z^{0.019a}}\right)^{1 + 1.8/Z_N}, \quad a = \upsilon Z_N^{-0.52} \tag{3.32}$$

for *solid* targets where Z denotes the *target* atomic number.

In Fig. 3.6, experimental equilibrium CSD in solid (left) and gas (right) targets are shown in comparison with results obtained by different formulae. Experimental data on CSD of krypton ions in collisions with C foil at 6 MeV/u are in better agreement with the Shima formula (3.27). The CSD data of Mg ion beam passing through hydrogen gas (right figure) are well described by the Gaussian functions, (3.21).

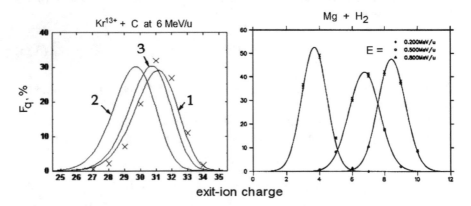

Fig. 3.6 Equilibrium charge-state distributions (CSD) in solid and gas targets. Left: CSD of krypton ions in collisions with C foil at 6 MeV/u. Crosses—experiment [205], curves—result of using different semiempirical formulae: 1—Shima [200], 2—Schiwietz [202], 3—To [204]. Right: CSD of Mg ion beam passing through hydrogen gas at different energies indicated. Symbols—experiment [203], solid curves—the Gaussian distribution fits

3.6 Difference Between Effective and Mean Charges of Fast Heavy Ions in Solids

Effective Z_{eff} and equilibrium mean $\bar{q}$ charges of fast ions are introduced in (2.11) and (3.6), respectively. The effective charge Z_{eff} is expressed via ratio of the measured ion stopping power to SP of the reference ion, and $\bar{q}$ is a state of charge equilibrium, determined by a competition of electron-loss and capture processes. Both charges are used in many applications, as well as to clarify one of the most important questions— what is the equilibrium mean charge $\bar{q}$ of the projectile in a solid target ? The $\bar{q}$ value can not directly be measured *inside* the matter but only the mean charge $\bar{q}_{exit}$ *after* emerging from a solid.

The question about difference between effective Z_{eff} and the mean $\bar{q}$ charges is considered in many papers. Recently, the authors of [149] found a relationship between the charge state inside the solid and the effective charge and explained a seeming contradiction between these values by performing non-linear stopping power calculations, which demonstrated consistency between the charge state of ions in solids and the empirical effective charge values. It was concluded that in the realistic approximation, the value of the ion charge within the solid $\bar{q} \approx q_{exit}$, i.e., to the mean ion charge after emerging from a solid.

As has been mentioned before (see (2.11), the effective charge Z_{eff} is defined experimentally through the ratio of stopping powers dE/dx as

$$Z_{eff}^2 = Z_0^2 \, \frac{dE/dx(\upsilon, Z_N)}{dE/dx(\upsilon, Z_0)}, \tag{3.33}$$

where Z_N denotes the nuclear charge of the projectile ion and Z_0 of the reference ion. This relation follows from the results of the perturbation theory according to which dE/dx is proportional to Z_N^2 provided the Bohr criterion (Sect. 3.4)

$$Z_N^2/\upsilon \ll 1, \tag{3.34}$$

is fulfilled where υ is the projectile velocity. Under this condition, it is adopted that

$$\bar{q}_{exit} > Z_{eff}. \tag{3.35}$$

However, the condition (3.34) is violated in many cases, for example if $\upsilon = 14$ a.u. ($E \approx 5\,\text{MeV/u}$) and $Z_N > 14$, i.e., if the projectile is heavier than silicon. Therefore, the quantity Z_{eff} defined through the SP is not a very good parameter (see [14]).

In the work [149] the SPs are calculated in the non-linear approximation for projectiles with $1 < Z_N < 92$ and $Z_0 = 1$ at energies of 0.1–100 MeV/u. The SPs values are found with a help of the transport cross sections, and the phase shifts by numerical integration of the radial Schrödinger equation for the projectile moving in a carbon media. As an input values of the projectile charge $q(\upsilon)$ inside the solid, the $\bar{q}_{exit}$, calculated by Nikolaev (3.26) and Schiwietz (3.32) formulae are used. The

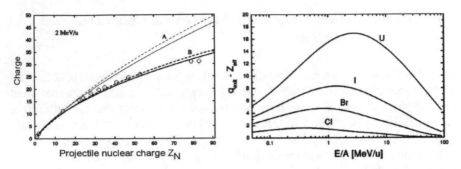

Fig. 3.7 Left: Mean and effective charges of heavy ions colliding with C foil at energy of 2 MeV/u as a function of the projectile nuclear charge Z_N. Curves A: Mean charges given by Nikolaev-Dmitriev formula (3.26), dashed curve, and Schiwietz formula (3.32), solid curve, used as input values for non-linear calculations of effective charges represented by curves B. The circles are experimental data of effective charges obtained from energy loss measurements [206]. Right: Differences between the mean charge of ions emerging from solid targets q_{exit}, using the Schiwietz formula (3.32), and the effective charges Z_{eff} using the non-linear calculations for Cl, Br, I and U ions as a function of the ion energy. From [149]

results of calculated mean and effective charges of projectile ions are presented in Fig. 3.7.

In Fig. 3.7, left, two sets of calculated charges of projectiles with the nuclear charge $1 < Z_N < 92$ impinging the carbon target are displayed (curves A and B) in comparison with effective charges obtained from the energy-loss measurements (circles). Curves A represent $\bar{q}_{exit}$ values calculated by Nikolaev-Dmitriev formula (3.26), dashed curve, and Schiwietz formula (3.32), solid curve. Effective charges calculated this way are shown by curves B which are in a good agreement with experimental data [206].

As seen from Fig. 3.7, left, the curves A and B diverge at $Z_N > 20$, i.e., for projectiles heavier than calcium. The difference between $\bar{q}_{exit}$, calculated by Schiwietz formula (3.32), and Z_{eff} by non-perturbative method in for Cl, Br, I and U projectiles in C foil as a function of ion energy is shown in Fig. 3.7, right. The largest deviation between $\bar{q}_{exit}$ and Z_{eff} is seen for $E \approx 1$–6 MeV/u. On the basis of results obtained in [149], it was concluded that the ion charge within the solid is reasonably well represented by the emerging ion charge, i.e., $\bar{q} \approx \bar{q}_{exit}$.

3.7 Computer Codes for Calculating Evolutions of Charge-State Distributions (CSD)

At present, there are a few computer codes for calculation of charge-state distributions, i.e., evolutions of non-equilibrium and equilibrium fractions, for heavy many-electron ions passing through media, mainly gaseous and solid targets. Below, a brief description of these codes is given.

3.7.1   ETACHA Code

ETACHA (old version) [207] is one of the first computer codes created for calculation of the evolution of ion charge-state fractions in *foils* at energies $E = 10-80\,\mathrm{MeV/u}$, corresponding to the working diapason of the GANIL accelerator, (Grand Accélérateur National d'Ions Lourds in Caen, France). The code is based on solving 84 differential balance rate equations (3.1) and (3.2) for projectiles with 28 electrons at maximum (up to Ni-like ions) with 1s, 2s, ..., 3d electron shells.

The code calculates the cross sections, required for solving the balance equations, in the 1st-order perturbation approximation and its modifications (PWBA, CDW, eikonal) for electron loss and capture (NRC and REC), excitation and de-excitation cross sections, radiative and Auger transition probabilities with account for the kinetic energy loss of the incident ions. Some cross sections can be enter into the input data by the user. The program is very useful and quite easy to use. However, comparison with experimental data shows that an applicability of ETACHA code is limited in the case of heavy projectiles with the nuclear charge $Z_N > 18$ and high energies $E > 30\,\mathrm{MeV/u}$ (see, e.g., [205, 208, 209]).

Figure 3.8 shows a comparison of experimental data of charge-state fractions of C^{2+} (light) and Ar^{8+} (heavy) ions in carbon foil at energies 4.3 and 6 MeV/u, respectively, with calculations by ETACHA code [207]. In the case of C^{2+} ions, there is quite good agreement between experiment and theory, meanwhile, for heavy ions, Ar^{8+}, the discrepancy is large. As is seen from the figure, the general behaviour seems to be similar both in the observed and calculated charge-state fraction distributions but

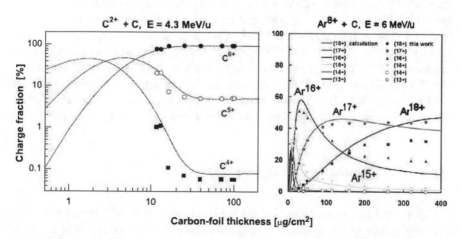

Fig. 3.8 Left: Charge-state fractions as a function of the carbon-foil thickness for 4.3 MeV/u C^{2+} projectile ions: symbols experimental data, solid curves results of ETACHA code [207]. From [209]. Right: Charge-state fractions as a function of the carbon-foil thickness for the 6.0 MeV/u Ar^{8+} projectile ions: symbols experimental data, solid curves result of ETACHA code [207]. From [208]

the charge-state distributions are in a serious disagreement with each other at large thicknesses corresponding to the charge equilibrium. This disagreement between experimental data and results by ETACHA code is increased with the projectile atomic number increasing. The authors of [208, 209] concluded that the main reason for such discrepancies can arise from the fact that the ETACHA code does not take properly into account the *target-density* effect in the solid targets which is expected to be very large for heavy ions and and many-electron targets (see Sects. 4.3 and 6.6). ETACHA old code can be downloaded from GSI website [210].

A description of a new version of ETACHA code is presented in [212]. It is extended towards lower velocities, covering approximately an energy range $E = 0.05–0.5$ MeV/u, and heavier ions with up to 60 electrons in $n = 1–4$ shells. For this purpose, five versions of a new ETACHA code (ETACHA23, ETACHA3, ETACHA34, ETACHA4, ETACHA45, depending on up to which shells are taken into account) were developed on the basis of non-perturbative approximations for the cross sections such as CDW-EIS (Continuum Distorted Wave Eikonal Initial State), and SEIK (Symmetric Eikonal) models, and increasing the total number of the projectile electronic configurations from 84 up to 1283. Comparisons of evolutions calculated by new ETACHA4 subroutine for selected charge-states of 13.6 MeV/u Ar^{10+} in carbon foil with ETACHA old is presented in [212], and with the BREIT code (Sect. 3.7.3) in Fig. 3.12.

As a demonstration for lower-energy and more-electronic configurations, charge-state evolutions for 2.0-MeV/u S^{q+} ions ($q = 6–16$) and 1.0-MeV/u C^{q+} ions ($q = 3–6$) after C-foil penetration reproduced by ETACHA4 and ETACHA3 versions, respectively, are shown in Fig. 3.9a, b, as well as charge-state distributions for 1.0-MeV/u W^{30+} ions after C-foils of 4.6, 9.9, and 20 μg/cm^2 reproduced by ETACHA45 and ETACHA4 versions in Fig. 3.9c. It is seen that equilibrium mean charge for 2.0-MeV/u S ions by ETACHA4 was 12.63, which reproduced the experimental value 12.7 within 0.5%, and that ETACHA4 qualitatively reproduced the *quasi-equilibrium* described in Sect. 3.1. Experimental charge-state distributions for W^{30+} initial ions after 9.9 and 20 μg/cm^2 C-foils almost coincided, showing an equilibrium thickness is 9.9 μg/cm^2, which ETACHA45 successfully reproduced, although calculated equilibrium mean charge was higher and distribution width wider than the experiment.

3.7.2 CHARGE and GLOBAL Codes

CHARGE and GLOBAL codes [117] have been developed for calculation of evolutions of the charge-state fractions for projectiles at high and relativistic energies, $30 < E < 2000$ MeV/u, in gaseous and solid targets with atomic numbers $1 \leq Z_T \leq 92$. The CHARGE and GLOBAL codes are available online at GSI website [210] (see also [211] for descriptions of the codes). The code CHARGE is based on the Allison three-component model (see [186] and Sect. 3.2), and is applied to describe three largest charge-state fractions, for example, bare, H- and He-like ions at relativistic collisions.

Fig. 3.9 a Mean charge-state evolutions for 2.0-MeV/u S^{q+} ions ($q = 6$–16) after C-foil penetration calculated by ETACHA4 code. **b** Charge-state evolutions for 1.0-MeV/u C^{q+} projectile ions ($q = 3$–6) after C-foil penetration. Circles, squares, triangles, and diamonds denote experimental fractions for C^{6+}, C^{5+}, C^{4+}, and C^{3+} outgoing ions, respectively, whereas full lines are calculated fractions by ETACHA3 code. **c** Charge-state distributions for 1.0-MeV/u W^{30+} ions after C-foils of 4.6, 9.9, and 20 μg/cm². Symbols denote experimental results and calculated distributions by ETACHA45 and ETACHA4 codes are drawn with full and dashed lines, respectively

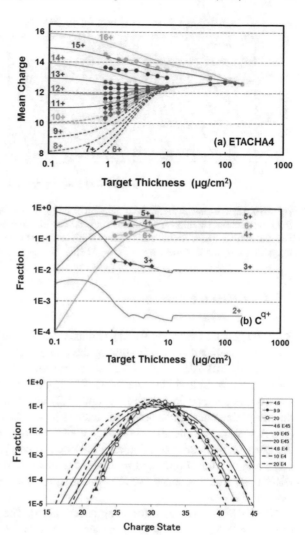

GLOBAL is used for projectiles with a nuclear charges $Z_N > 30$. The main advantages of the GLOBAL compared to CHARGE are as follows:

1. GLOBAL can calculate up to 28 fractions instead of 3 in the CHARGE code.
2. GLOBAL takes into account the energy loss of projectiles in matter.

In the case of three charge states, results of both codes almost coincide. The accuracy of the charge-state fraction calculations by the CHARGE and GLOBAL codes is within a factor of 2. Figure 3.10 shows experimental charge-state evolution of 955-MeV/u uranium ions passing through thin Al, Ag, and Au foils as a function of the target thickness in comparison with the GLOBAL calculations. Charge-states distributions of heavy ions in foils at lower energies, $18\,\text{MeV/u} < E < 44\,\text{MeV/u}$, are given in [213].

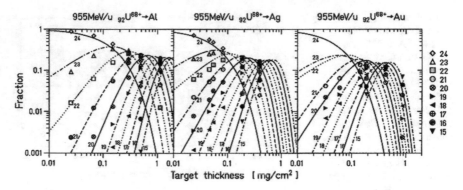

Fig. 3.10 Charge-state evolution as a function of target thickness for 955-MeV/u U^{68+} ions (with 24 initial electrons) in thin Al, Ag, and Au targets. The lines represent the predictions of GLOBAL. The numbers written next to the lines denote the number of electrons in the projectile. The curves are predicted by the GLOBAL code. From [117]

3.7.3 BREIT Code

Recently [214], a new code BREIT (Balance Rate Equations for Ion Transportation) has been developed for calculation of the charge-state fractions $F_q(x)$ for any ions with energy $50\,keV/u \leq E \leq 50\,GeV/u$ and number of fractions $3 \leq N_F \leq 200$ for gaseous, solid and plasma targets. The code is based on numerical solution of a set of the balance equations (3.1) and (3.2) using the diagonalization method of the interaction matrix, consisting of electron-loss and electron-capture cross sections (see Appendix A in detail). Unlike the codes mentioned above, in the BREIT code the cross sections are not calculated but need to be given in a input file, including cross sections for single- and *multiple-electron* processes. Therefore, the users of BREIT are free to choose the charge-changing cross sections from any theory or experiment.

The BREIT is available for using online [215]. The web interface allows one to use the code from any platform via web browsers. The results are made available in the text, PDF and interactive figures formats. The description of the code and examples of the input files are available on websites [216, 217], respectively.

For calculating the charge-state fractions in *plasma* targets, the input file into the BREIT should contain information about rate coefficients $N\upsilon\sigma$ with account for additional processes occurring in plasmas—radiative and dielectronic recombinations, ionization by plasma particles with the density N (see [42, 43] and Chap. 7).

Typical example of the BREIT output is shown in Fig. 3.11 for collisions of $1.4\,MeV/u$ U^{4+} ions with molecular hydrogen target at 20 mbar pressure. There, the calculated evolutions (left figure) and equilibrium fractions (right figure) of 37 charge states are shown for $4 \leq q \leq 40$, using 104 input single- and multiple-electron loss

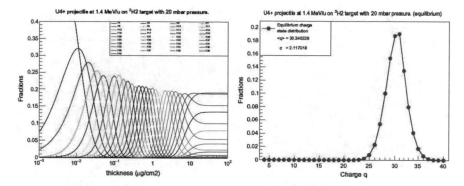

Fig. 3.11 Left: An example of the BREIT output figure of the non-equilibrium charge-state fractions $F_q(x)$ as a function of molecular hydrogen thickness for charge states $4 \leq q \leq 40$ of the 1.4 MeV/u incident U^{4+} ions at 20 mbar gas pressure. Right: An example of the BREIT output figure of the *equilibrium* charge-state fractions F_q as a function of the uranium charge-state. This example corresponds to the asymptotic (equilibrium) limit of the charge-state fractions of the left figure distributed along the charge axis. The mean charge $< q >$ and the standard deviation σ of the Gaussian distribution are given in the left upper corner of the right figure. From [214]

and capture cross sections calculated with account for the target density effect (see [214] for detail). A Gaussian distribution of equilibrium charge-state fractions is given in the right figure also showing the equilibrium mean charge $\bar{q}$ and a standard deviation σ.

Calculations of non-equilibrium and equilibrium charge-state fractions, performed by the BREIT code, are compared with experimental data as well as results of ETACHA (new) and GLOBAL codes in Fig. 3.12. The target-density effect is important in these cases and was taken into account in the BREIT input for loss and capture cross sections using the results of [182]. The losses of ion kinetic energy, obtained experimentally or calculated by the SRIM code [127], do not exceed 10% of the incident ion energy.

The evolution of the largest charge states of argon ions in collisions of Ar^{10+} ions with a carbon foil at 13.6 MeV/u is shown in Fig. 3.12, left. Experimental data, recorded at GANIL, France, are denoted by symbols with error bars. Calculations by the new ETACHA code [212] are shown by dashed curves and the BREIT results by solid curves. Charge-state fractions $F_{16}(x)$, $F_{17}(x)$, and $F_{18}(x)$, calculated by both codes, are in good agreement with experimental data. The BREIT results give the equilibrium thickness for this case: $x_{eq} \approx 1000 \, \mu g/cm^2$.

In Fig. 3.12, right, comparison of experimental evolution [117] of various charge-state fractions with theoretical predictions is given for Ne-like Au^{69+} ions in a gold foil at 1 GeV/u energy. Dashed curves represent the results of the GLOBAL code [117] and the solid curves the BREIT results. Calculated results for the most intense fraction F_{79} are in agreement with experiment and with each other within a factor of 2.

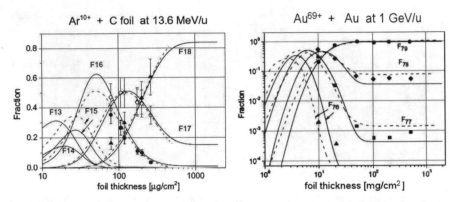

Fig. 3.12 Left: Charge-state fractions of argon ions in collisions of 13.6 MeV/u Ar^{10+} ions with carbon foil as a function of a foil thickness. Experimental data [212] are denoted by symbols with error bars, the dashed curves are the new ETACHA code result [212] and solid curves are the BREIT results with 16 input cross sections accounting for the density effect. Right: Charge-state evolutions of 1000-MeV/u Ne-like Au ions downstream Au foil. Symbols experimental data [117], dashed curves the GLOBAL code [117], solid curves the BREIT result with 50 input cross sections. From [214]

3.7.4 Other Codes

One has to mention other computer codes and methods used for calculation of the charge-state fractions but not so often used than those mentioned above. Recently, a Monte-Carlo (MC) code was developed in [173, 188, 218] to calculate charge-state fractions of fast heavy ions colliding at energies $E > 10$ MeV/u with *plasma* targets. The MC code uses the ETACHA old version [207] for calculation of the input cross sections: NRC and REC cross sections, ionization and excitation of projectiles, as well as de-excitation and recombination cross sections. Similar to ETACHA old code, the MC code can calculate the fractions of ions with the number of electrons less than 28 for penetration through plasmas with ion densities $10^{18} < N_i < 10^{23}$ cm^{-3} and temperatures $10 < T_i < 200$ eV. The density effect is taken into account in the MC code.

In practice, various methods are also used for solving the balance rate equations, for example, a matrix method [219], a semiempirical method [203, 220] and others.

Finally, concerning calculations of the charge-state fractions on the basis of the differential balance equations (3.1) and (3.2) with the cross sections as coefficients, we have to stress the following:

1. since the accuracy of determination of experimental and theoretical cross sections lies in the limits of 10–50%, the accuracy of calculated charge-state fractions $F_q(x)$ is about a factor of 2, especially concerning small charge-state fractions. The situation with *equilibrium* fractions $F_q(\infty)$ is a little better because they are defined mainly through the *ratios* of one-electron loss-to-capture cross sections

(see (3.17)) which can be determined with higher accuracy than the multi-electron cross sections,

2. in a dense target (dense gas or plasma, foil) charge-changing cross sections, in principle, depend on the target thickness x that should be taken into account. The dependence on x includes two main factors: the target-density effect and the loss of the projectile kinetic energy since cross sections depend on the beam energy,

3. in the case of heavy projectile colliding with many-electron targets, an influence of multi-electron processes can be very large (20–30%), therefore, such multiple-electron processes, if possible, should be included into calculations.

Chapter 4
Electron Capture Processes

Abstract This chapter is devoted to consideration of electron capture (EC), which is one of the most important recombination mechanisms in collisions of ions with various targets. Special features of EC are described such as the presence different atomic particles before and after collision, a preferential capture of inner-shell target electrons at high energies, the role of excited states of the projectile ion after EC, different scaling laws at low and high energies and so on. The influence of the target-density effect as well as the use of the Bragg's additivity rule for EC cross sections are described. Information on multiple-electron capture, which is very important for many applications, is also given. EC processes at very low energies, where the *molecular* effects play a key role, are considered in detail in Chap. 5.

4.1 Main Properties of Electron-Capture Processes

This section is devoted to the consideration of single-electron capture processes arising in collisions of fast heavy projectiles with atomic targets:

$$X^{q+} + A \rightarrow X^{(q-1)+} + A^{+}, \tag{4.1}$$

where X^{q+} and A denote the projectile ion with charge q and the target atom or molecule, respectively.

Usually, single-electron capture is a dominant process, i.e., gives the main contribution to the total capture cross section, but in the case of heavy projectile ions, multiple-electron capture becomes very important and contributes up to more than 50% to the total capture cross section because of a strong long-range Coulomb field of the projectile makes it possible to capture a few target electrons simultaneously with a large probability.

Cross sections of reaction (4.1) can be very large, $\sim 10^{-14} - 10^{-16}$ cm^2, especially at low and intermediate energies, hence, these processes play a significant role in atomic physics, astrophysics, plasma and accelerator physics.

Reaction (4.1) strongly influences the charge-state distributions in ion beams passing through solid, gaseous and plasma targets [1], and plays a key role in

© Springer International Publishing AG 2018
I. Tolstikhina et al., *Basic Atomic Interactions of Accelerated Heavy Ions in Matter*, Springer Series on Atomic, Optical, and Plasma Physics 98,
https://doi.org/10.1007/978-3-319-74992-1_4

thermonuclear plasma confinement in tokamaks [32, 221] creation of an inversion population in a short-wavelength spectral range [222, 223]. In particular, reaction (4.1) is an effective mechanism of excitation transfer in plasmas [25–27], in astrophysics [48], in accelerator techniques [45, 224], and other areas.

The main competitive process to electron capture is electron loss—ionization of a projectile by a target atom or molecule (see Chap. 6). Both processes have different dependencies on relative collision velocity v, projectile ion charge q, and atomic structure of the target atom and, therefore, the contribution of each process strongly depends on the considered ion energy range (see Fig. 1.2).

As electron capture process (4.1) is a complicated rearrangement reaction with different particles before and after collision; therefore, theoretical investigation of these reactions, even at high collision energies, constitutes a much more difficult problem than that of *electron-ion* collisions. If the cross sections for electron-ion collisions can be calculated quite accurately, with an accuracy of 10–20%, to describe the capture reactions with an accuracy of a factor of 2 is a rather tedious task even in a high-energy region. This is due to the great difficulties arising in a description of these processes: the use of different interaction potentials before and after collision (the so-called *post-prior discrepancy*), the non-orthogonality of the wave functions of the system in the initial and the final channels, the Coulomb interaction between two ions in the final channel and its absence in a collision of an ion with an atom in the initial channel, and so on.

In practice, approximate methods are applied to electron capture depending on the atomic parameters: relative velocity v of the colliding particles, their atomic structures, and the *resonance defect* ΔE of the reaction, i.e., the difference between binding energies of the active electron in the target atom I_A before collision and the resulting ion I_X after collision (Fig. 4.1):

$$\Delta E = I_A(n_0 l_0) - I_X(n_1 l_1), \quad I_A > 0, \quad I_X > 0, \tag{4.2}$$

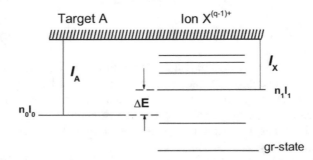

Fig. 4.1 Scheme of the target atom A, where the active electron was in the state $n_0 l_0$ with the binding energy $I_A(n_0 l_0)$ before capture reaction, and the ion $X^{(q-1)+}$ after reaction with the captured electron in the state $n_1 l_1$ and the binding energy $I_X(n_1 l_1)$. $n_0 l_0$ and $n_1 l_1$ the principal and orbital quantum numbers of the target atom and the resulting ion, respectively, *gr-state* denotes the ground state of the $X^{(q-1)+}$ ion, and ΔE the resonance defect. From [225]

where $n_0 l_0$ and $n_1 l_1$ denote the principal and orbital quantum numbers of the target atom A and the resulting ion $X^{(q-1)+}$, respectively. We note that the resonance defect ΔE can be both positive and negative.

Among the different theoretical methods of calculating electron-capture cross sections, one can mention a few basic ones which give a satisfactory description of the experimental data: the close-coupling method with an atomic or molecular basis [226–228], the electron tunneling (through the Coulomb barrier) model [229], the absorbing sphere model based on the Landau- Zener theory [230], the classical over-barrier-transition model [231], the distorted-wave approximation with normalization [232, 233], relativistic treatment based on solving the two-center Dirac equation for the colliding system 'nucleus + H-like target' [234], and others. Most of the methods mentioned are described in [23–42]. Methods for calculating electron-capture cross sections at low-energy collisions (for example, the adiabatic approximation, the ARSENY code) are described in detail in Chap. 5.

At different ion-atom collision energies, electron capture occurs as a result of different physical processes, and two main ranges of relative collision velocity v are usually defined: the *adiabatic* region with $v < v_e$, and the *non-adiabatic* one with $v > v_e$, where v_e denotes an electron orbital velocity of the target atom. At low-energy collisions $v < v_e$, the target bound electrons adiabatically react to the varying field of the moving incident ion, and, thus, a *quasi-molecular* treatment is applied, when the solution of the problem is based on the expansion of the total wave function of the system in terms of the quasi-molecular wave functions at fixed internuclear distance R, and transitions between different states proceed as occurring between quasi-molecular potential terms corresponding to localization of the active electron close to one of the nuclei. This treatment is especially effective for describing the resonance ($\Delta E = 0$) and quasi-resonance ($\Delta E \approx 0$) electron capture (see, e.g., [19, 28, 36]).

For collision energies $E > 25 \text{keV/u}$ ($v > 1 \text{a.u.}$), when the impact (projectile) velocity is higher than the target-electron orbital velocity, $v > v_e$, the non-resonant electron capture prevails and the quasi-molecular method is not valid. This is mainly related to an influence of the momentum transfer carried away by the captured electron, the *translation factor* $\exp(i \mathbf{v} \cdot \mathbf{r})$, which can be neglected at low ion velocities, allowing one to present the interaction matrix elements through a splitting of the corresponding molecular terms (see [235]).

At intermediate energies $E \sim 1$–25keV/u, the target outer-most electrons are captured by ions with high probabilities, and due to a contribution of electron capture into a large number of excited n-states of the $X^{(q-1)+}(n)$ ion, the total electron-capture cross section has a quasi-constant character, i.e., its magnitude is nearly independent of the collision energy. The quasi-constant behavior of the electron-capture cross sections in collisions of highly charged ions with neutral atoms was predicted in the paper [236]. This quasi-constant magnitude of the capture cross section, which is closest to experimental data, was estimated in the model of *electron tunneling* through the Coulomb barrier created by the target atom and the projectile as [229]:

$$\sigma(\upsilon) \approx \text{constant} \approx 10^{-15} \frac{q}{(I_T/Ry)^{3/2}} \, [\text{cm}^2], \quad q \geq 5, \quad \upsilon^2 < I_T/Ry, \quad (4.3)$$

where I_T/Ry denotes the ionization potential of the target atom in Ry units
(1 $Ry \approx 13.606$ eV). At intermediate collision energies, the electron capture leads
to a preferential population of excited n-states of the resulting ion $X^{(q-1)+}(n)$ given
as:

$$n \approx q^{3/4} \left(\frac{I_T}{Ry} \right)^{-1/2}. \quad (4.4)$$

If the ion velocity υ exceeds the orbital velocity of the target electron, electron
capture processes can be described in the first-order perturbation theory (or its modifications) on interaction of the active electron with the projectile, e.g., the distorted-
wave approximation or the Brinkman-Kramers approximation with a multichannel
normalization in the impact-parameter representation (see Sect. 4.2).

At relatively high, but non-relativistic, energies corresponding to $E = 25$ keV/u–
30 MeV/u, EC is characterized by the preferential capture of *inner-shell* target electrons, as the electron-shell structure of the target atom becomes substantial. The
preferential role of the inner-shell target electrons is the main property of the capture reactions, which makes it different from other processes in collisions of fast ions
with atoms.

Let us consider an electron capture from the target shell a with the binding energy I_a and the orbital velocity $\upsilon_a \sim I_a^{1/2}$. The corresponding cross section has its
maximum when the ion velocity υ is close to υ_a, $\upsilon \approx \upsilon_a$, i.e., when the so-called
velocity matching takes place. At $\upsilon > \upsilon_a$ the capture cross section falls off as

$$\sigma_{EC} \sim q^5 Z_T^5/\upsilon^{11} \sim E^{-5.5}. \quad (4.5)$$

With the energy increasing, the velocity matching takes place for another electron of
the deeper shell b having orbital velocity $\upsilon_b > \upsilon_a$, $I_b > I_a$, the capture cross section
will have a local maximum, and so on until a capture of the target 1s electrons
will occur. As a result, the total capture cross section $\sigma_{EC}^{(tot)}$, summed over all target
electrons, decreases much more slowly than by the $E^{-5.5}$ law valid only at very high
energies, where the cross section is mainly defined by the capture of the deepest 1 s
electrons. Due to capture of *all* target electrons, the cross sections for collisions with
light targets (H, He) decrease much faster than those with heavy ones (Ne, Ar, Kr,
Xe) having several electron shells.

The effect of the preferential capture of inner-shell target electrons is illustrated
in Fig. 4.2 for collisions of protons and U^{42+} ions with Ar target. As is seen, at high
enough energies, capture of inner-shell electrons gives the main contribution to the
total cross sections, while a contribution of outer-most electrons of Ar atoms is very
small due to the asymptotic law (4.5).

As has been discussed in Sect. 1.4, at high energies $E > 200$ MeV/u, radiative
electron capture accompanied by a photon radiation, begins to play a major role in
recombination processes; REC processes are considered in Sect. 4.6.

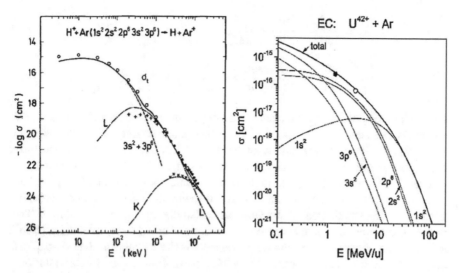

Fig. 4.2 Left: Electron-capture cross section between protons and argon atoms. Symbols— experiment: open [237] and solid [238] circles—total cross sections, x and + correspond to capture from L and M electrons of Ar - [238, 239], respectively. Theory: broken and solid curves— contribution from electron capture from different shells of Ar [225]. From [225]. Right: Single-electron capture cross sections of U^{42+} ions colliding with Ar target. Experiment: solid circle [95], open circle [103]. Solid curves: calculations by the CAPTURE code using the normalized Brinkman-Kramers approximation. Contributions from different shells of Ar are shown as well as the total cross section summed over capture of all Ar electrons. From [235]

4.2   Methods and Computer Codes for Calculating Single-Electron Capture Cross Sections

At present, there are several methods and computer codes often used for calculating single-electron capture cross sections in collisions of heavy many-electron ions with atoms: CTMC (classical trajectory Monte Carlo) method for energies $E > 1\,\text{MeV/u}$, CDW (Continuum Distorted Wave) approximation for $E > 10\,\text{MeV/u}$, and the normalized Brinkman-Kramers approximation for $E > 10\,\text{keV/u}$. The accuracy of these calculations is within a factor of 2. These methods and corresponding computer codes are briefly discussed here.

The CTMC method [240] (and corresponding computer code) is a based on the numerical solution of a system of the Hamilton classical-motion equations for all projectile and target electrons using a large number of impact parameters $\sim$5000 for the particle trajectories. The 'ion + atom' system consists of $6(N+2)$ nonlinear first-order equations in partial derivatives and is solved numerically for the coordinates and momenta for all N electrons and two nuclei of the system. Thus, for one electron moving in the Coulomb field of two nuclei a and b, the classical Hamilton equations can be written as

$$H = \frac{p_a^2}{2M_a} + \frac{p_b^2}{2M_b} + \frac{p_e^2}{2M_e} + \frac{Z_a Z_b}{R_{ab}} + \frac{Z_a e}{R_{ae}} + \frac{Z_b e}{R_{be}}, \tag{4.6}$$

$$\frac{\partial C_j}{\partial t} = \frac{\partial H}{\partial p_j}, \quad \frac{\partial p_j}{\partial t} = \frac{\partial H}{\partial C_j}, \quad j = x, y, z, \tag{4.7}$$

where C_j and p_j denote coordinates and momenta of an electron and nuclei, m_e and e electron mass and charge, $M_{a,b}$ and $Z_{a,b}$ masses and charges of the nuclei a and b, and R_{ab} the distance between the nuclei. From (4.7) one has 18 bound first-order differential equations for coordinate and momentum evolutions of all particles.

The CTMC method is applied to the intermediate collision energy range, where *molecular* effects can be neglected. The use of the method is quite complicated, but electron-loss and excitation cross sections are also obtained in the calculation. This method is quite complicated because many electrons and atomic trajectories should be taken into account to get enough statistics for the calculated electron-capture, electron-loss and excitation cross sections. We note that because of the computational difficulties mentioned, the number of publications on CTMC electron-capture cross sections involving heavy ions is quite sparse (see, e.g., [241–243]).

The CDW method [233] is based on the modified Born (distorted-wave) approximation at sufficiently high energies $E > 10$ MeV/u. This method utilizes the Clementi-Roetti functions as the bound-state wave functions, and the Coulomb functions for continuum states.

The eikonal approximation [244] is used for computing one-electron capture cross sections and is based on the semi-classical approximation with three main assumptions: a straight ion trajectory, hydrogen-like bound electron wave functions, and distorted wave functions in the final channel described by the eikonal phase factor (see, e.g., [241]).

The CAPTURE code [245] is intended for calculating probabilities and cross sections of one-electron capture and is based on the Brinkman-Kramers approximation [246]. This code calculates the normalized electron-capture probabilities $P^N(b, \upsilon)$ as functions of the impact parameter b and collision velocity υ, as well as single-electron capture cross sections into excited n-states of the resulting ion $X^{(q-1)+}(n)$ and the total (summed over n) cross sections $\sigma_{tot}(\upsilon)$:

$$\sigma_{tot}(\upsilon) = \sum_{n=n_0}^{n_{cut}} \sigma_n(\upsilon), \quad \sigma_n(\upsilon) = \sum_s \sigma_{sn}(\upsilon), \tag{4.8}$$

$$\sum_s \sigma_{sn}(\upsilon) = 2\pi \int_0^\infty P_{sn}^{(N)}(b, \upsilon) \, b \, db, \tag{4.9}$$

$$P_{sn}^{(N)}(b, \upsilon) = \frac{P_{sn}(b, \upsilon)}{1 + \sum_{n'=n_0}^{n_{max}} P_{sn'}(b, \upsilon)}. \tag{4.10}$$

Here, $P_{sn}(b, \upsilon)$ denotes the Brinkman-Kramers probability of a capture from the target electron shell s into the n-state of the $X^{(q-1)+}(n)$ ion, and n_{max} is the maximum principal quantum number accounted for in the CAPTURE code. The parameter n_{cut} depends on the target density: n_{cut} is large for low-density targets (rarefied gases) and significantly decreases in the case of high-density targets (foils) due to the influence of target-density effect (see Sect. 4.3).

In the CAPTURE code for the *active* electron (electron to be transferred), the hydrogen-like wave functions are employed in the initial (the target atom) and final states, with the effective charge accounting for the screening of non-active electrons. The main advantage of the CAPTURE code is the use of the *normalized* capture probabilities which are always less than unity: $P^{(N)}(b, \upsilon) < 1$. This circumstance allows one to perform calculations over a wide energy range: from a few tens keV/u to a few tens MeV/u. We note that the CAPTURE code can provide calculations with account for the final states with a large number n (up to $n_{max} \sim 500$), which requires for calculation of the capture cross sections for highly charged projectile ions, $q \gg 1$, e.g., U^{92+} projectile.

For rough estimation of single-electron capture cross sections, a semiempirical formula introduced by Schlachter et al. [247] is often used:

$$\sigma_{Sch} = 1.1 \cdot 10^{-8} \ [\text{cm}^2/\text{atom}]$$
$$\times \frac{q^{0.5}}{Z_T^{1.8} \tilde{E}^{4.8}} \left[1 - \exp(-0.037\tilde{E}^{2.2})\right] \cdot \left[1 - \exp(-2.44 \cdot 10^{-5}\tilde{E}^{2.6})\right], \quad (4.11)$$

$$\tilde{E} = E / \left(Z_T^{1.25} q^{0.7}\right), \quad q \geq 3, \quad 10 \leq \tilde{E} \leq 1000, \quad (4.12)$$

where q denotes the incident projectile charge, Z_T atomic number of the target, $\tilde{E}$ a reduced energy, and E the ion energy in keV/u. This formula is based on the experimental data available at that time with some general assumptions on the EC cross section behavior. The formula can be used for practical applications with accuracy of a factor of 2 over a wide energy range.

At low and high energy limits, the Schlachter semiempirical formula (4.11) has the following asymptotic behavior:

$$\sigma_{Sch}(E \to 0) \approx 10^{-14} \ [\text{cm}^2/\text{atom}] \times \frac{q^{0.5}}{Z_T^{1.8}}, \quad (4.13)$$

$$\sigma_{Sch}(E \to \infty) \approx 10^{-8} \ [\text{cm}^2/\text{atom}] \times \frac{q^{3.86} Z_T^{4.2}}{E^{4.8}}. \quad (4.14)$$

A more accurate estimation of EC cross sections at $E \to 0$ is given by (4.3). Because of the different dependencies of (4.13) and (4.3) on atomic parameters, the Schlachter formula should be used with caution at very low energies as it is known to over/under estimate experimental cross-section values.

In the middle-to-high energy range, a semiempirical formula by Knunsen et al. [248] is often used (Sect. 4.4), which is based on the extended Bohr-Lindhard model, for estimating single-electron capture cross sections for ions with initial charge states $q \geq 4$ from atomic target.

For low-energy collisions, a scaling formula by Müller and Salzborn [249] is used for highly charged initial ions, which is described in detail in Sect. 4.5. Recently, Imai et al. proposed a scaling formula for low-energy low-q projectile ions colliding with atomic and molecular targets [250], where the collisions become endothermic (or near-resonant) (see Chap. 5).

Typical examples for the EC cross sections, calculated by different methods, are presented in Fig. 4.3, left and right, in comparison with experimental data. Capture cross sections of H-like Ge^{31+} ions in Ne target are shown in Fig. 4.3, left, where experimental data are compared with results obtained with the CTMC, eikonal, and CDW approximations, as well as using the CAPTURE code and Schlachter semi-empirical formula (4.11). As is seen from the figure, all theoretical data are in rather good agreement with experiment. Electron-capture cross sections in collisions of lead ions with Ar target are displayed in Fig. 4.3, right. For collision energies $E >$ 3 MeV/u, both Schlachter semiempirical formula (4.11) and the CAPTURE result reproduce rather well reproduce experimental data but at lower energies, formula (4.11) is not valid due to the energy limitation given in the formula (4.12).

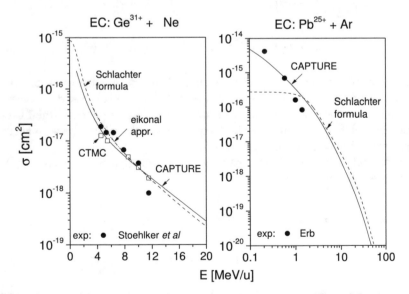

Fig. 4.3 Left: Single-electron capture cross sections in collisions of Ge^{31+} ions with Ne. Solid circles—experiment [97]. Theory: open squares—CTMC result [97]; dashed curve Schlachter empirical formula (4.11); the eikonal calculations [295] are practically identical with the values given by the Schlachter empirical formula; solid curve—CAPTURE code [235]. Right: Electron capture cross sections of Pb^{25+} with Ar. Experiment: solid circles [95]. Theory-same as in the left figure. From [97]

Molecular gaseous targets are of a special importance in interaction with residual-gas components in accelerator, stripping of heavy ions in diatomic gases etc., but even for 'simple' H_2 gas target an estimation of electron-capture cross sections meets severe difficulties because of a deviation from the Bragg's additivity rule (see [251–255] and Sect. 4.4).

4.3   The Target-Density Effect in Electron-Capture Processes

The density effect (DE) is highly important for electron-loss and capture processes in high-density targets. It drastically changes charge-changing cross sections and other related properties like equilibrium charge state $\bar{q}$, stopping powers in dense media and so on. The influence of the DE on electron-loss cross sections is considered in Sect. 6.6.

Let us consider an influence of the target density effect (DE) on one-electron capture cross sections for reaction:

$$X^{q+} + A \rightarrow X^{(q-1)+*}(n) + A^+, \qquad (4.15)$$

where A denotes a target atom or molecule, n the principal quantum number of the captured electron into the projectile ion in the final state, the asterisk stands for an excited state.

In the case of a low-density medium (a rarefied gas), there are only a few ion-atom collisions, and the $X^{(q-1)+}(n)$ ions are created in all possible quantum states n, which then decay into the ground state n_0 via radiative decay transitions. In this case, the total electron-capture cross section can be written in the form:

$$\sigma_{tot}(\upsilon) = \sum_{n=n_0}^{\infty} \sigma_n(\upsilon). \qquad (4.16)$$

In a high-density target, a collision frequency increases and the time interval between successive collisions becomes shorter than the lifetime of excited states so that the ions with the principal quantum numbers, higher than a certain number n_{max}, are ionized by the target particles in subsequent collisions, and ions with $n \leq n_{max}$ are stabilized via radiative transitions to the ground state. Then, the total electron-capture cross section with account for the density effects, σ_{tot}^{DE}, is defined by sum over n (4.16) but with the finite upper limit:

$$\sigma_{tot}^{DE}(\upsilon) = \sum_{n=n_0}^{n_{max}} \sigma_n(\upsilon), \quad n_{max} = n_0 + n_{cut}, \qquad (4.17)$$

where n_{cut} denotes the maximum principal quantum number of survived excited states. In the limiting case of very high target density $\rho_T \rightarrow \infty$, $n_{cut} \rightarrow 0$, and electron capture occurs only into the ground state:

$$\sigma_{tot}^{DE}(v) \approx \sigma_{n_0}(v). \tag{4.18}$$

This is a very important property of electron-capture processes that limits the average mean charges of ions penetrating various targets. Therefore, the number of 'surviving' $X^{(q-1)+}(n)$ ions decreases as the target density increases, resulting in a cross-section reduction.

The cut-off parameter n_{cut}, i.e., the maximum principal quantum number of non-ionized ions, can be estimated from a balance equation between ionization rate and the radiative decay probability of the excited state:

$$\rho_T v \sigma_{EL}(n_{cut}) = A(n_{cut}), \tag{4.19}$$

where ρ_T denotes the density of the target particles, σ_{EL} the electron-loss cross section of the $X^{(q-1)+}(n)$ ion by the target atoms and $A(n_{cut})$ the total radiative decay rate of the n_{cut} state. Using the classical Thomson formula for σ_{EL} and the classical Kramers formula for $A(n)$, one has [245]:

$$\sigma_{EL} \sim \frac{Z_T^2 n^2}{q^2 v^2}, \quad A(n) \sim \frac{q^4}{n^5}, \tag{4.20}$$

$$n_{cut} \approx q \left(\frac{10^{18}}{Z_T^2 \cdot \rho_T [\text{cm}^{-3}]} \right)^{1/7} \left(\frac{v^2}{10q^2} \right)^{1/14}$$

$$\approx q \left(\frac{10^{18}}{Z_T^2 \cdot \rho_T [\text{cm}^{-3}]} \right)^{1/7} \left(\frac{E[\text{keV/u}]}{250q^2} \right)^{1/14}, \tag{4.21}$$

where E is the projectile energy. As is seen, the target density effect is large (i.e., n_{cut} is small) when the density ρ_T and the target nuclear charge Z_T are large, and also when the projectile charge q and energy E are small. Although (4.21) is approximate, it exhibits the main dependencies of the cut-off value n_{cut} on the target density and other atomic parameters.

There is another formula for the n_{cut} parameter used for calculations, which is based on a *step ionization* of the scattered ion: the ion is first excited to the nearest states and then ionized by collisions with the target particles (see [255]). The corresponding formula has a more strict dependence on the atomic parameters:

$$n_{cut} \approx q \left(\frac{5 \cdot 10^{16}}{Z_T^2 \cdot \rho_T [\text{cm}^{-3}]} \right)^{1/9} \left(\frac{E[\text{keV/u}]}{q^6} \right)^{1/18}. \tag{4.22}$$

The influence of the density effect on the single-electron capture in collisions of 1.4 MeV/u U^{28+} ions with atomic hydrogen is demonstrated in Fig. 4.4a, b, where gas-pressure dependencies of n_{max} value and EC cross section are shown. As seen from the Fig. 4.4b, the density effect leads to a decrease by approximately two-thirds of EC cross sections in this case.

Fig. 4.4 Influence of the density effect on electron-capture process for collisions of 1.4 MeV/u U^{28+} ($n_0 = 5$) ground-state ions with atomic hydrogen. **a** Dependence of the n_{max} on the hydrogen pressure P, (4.22). **b** Gas-pressure dependence of EC cross section, (4.17) and (4.22). From [255]

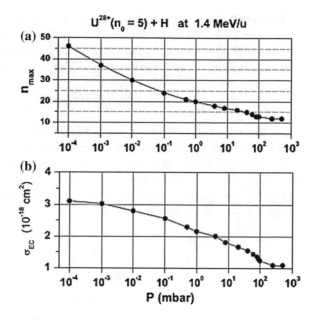

4.4 Bragg's Additivity Rule for Electron-Capture Cross Sections

The Bragg's additivity rule is used to interpret experimental data and to theoretically predict the atomic interaction data of *fast* ions with *molecular* targets. According to the rule, the projectile interaction with a molecule is presented as a sum of the interaction cross sections for its constituent atoms:

$$\sigma_{mol} = \sum_i n_i \sigma_i(Z_i), \qquad (4.23)$$

where n_i denotes the number of i-th atoms in the molecule and $\sigma_i(Z_i)$ the interaction cross section with an i-th atom of atomic number Z_i. For example, the electron-capture (or electron-loss) cross section of an ion colliding with H_2O molecule is presented as: $\sigma(H_2O) = 2\sigma(H) + \sigma(O)$. The Bragg's additivity rule is used because of the computational difficulties for molecular targets; however, this is partly justified because at high collision energies the main contribution to the electron-capture cross sections is made by the capture of inner-shell target electrons, which are approximately identical in atomic and molecular targets.

Quite often (but not always) the Bragg's additivity rule provides a reasonable agreement between theory and experiment depending on the projectile charge and velocity. For example, it works quite well for electron-loss processes, i.e., ionization of fast projectiles by molecules (Sect. 6.5) but for electron capture the situation is more complicated.

First of all, the Bragg's rule does not work for electron capture of ions colliding even with the simplest H_2 molecule. In [251, 252], the ratio $\sigma_{EC}(H_2)/\sigma_{EC}(H)$ of experimental single-electron capture cross sections in atomic and molecular hydrogen was analyzed and found not to be equal to 2 but changing in the range:

$$0.8 \leq \sigma_{EC}(H_2)/\sigma_{EC}(H) \leq 3.8, \tag{4.24}$$

depending on the incident ion charge and kinetic energy.

Experimental ratios $\sigma(H_2)/\sigma(H)$ are given in Fig. 4.5 as a function of a scaled ion energy showing that the ratio changes from 0.8 to 3.84 and equals to 2 only around the scaled energy $E(keV/u)/q^{4/7} \approx 40$ where q is the incident ion charge. Symbols in Fig. 4.5 correspond to experimental data and the dashed curve is an empirical fit, which is valid within 20% (see [251]):

$$\frac{\sigma(H_2)}{\sigma(H)} = 0.76, \quad X < 6,$$
$$= 1.76 + 0.0328(X - 6), \quad 6 < X < 100, \tag{4.25}$$
$$= 3.84, \quad X > 100,$$
$$X = E[\text{keV/u}]/q^{4/7}, \quad q \geq 5. \tag{4.26}$$

Equation (4.25) is valid for ions with rather high charge states $q \geq 5$ with the scaled energy, (4.26), taken from Bohr-Lindhard classical model for electron capture

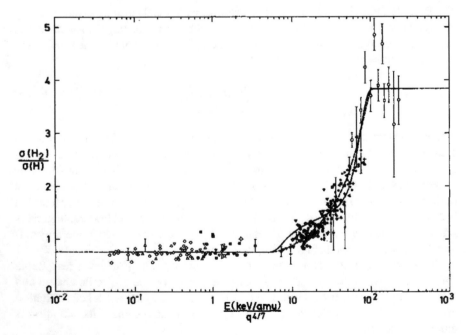

Fig. 4.5 Ratio of single-electron cross sections $\sigma_{EC}(H_2)/\sigma_{EC}(H)$ as a function of a scaled ion energy. Symbols—experimental data, the dashed curve—empirical fit, (4.25). From [251]

[180]. Possible reasons of deviation of the ratio from 2 are investigated theoretically in [230, 253] using the absorbing-sphere model.

Another semiempirical formulae for the $\sigma(H_2)/\sigma(H)$ ratio (within 30%) was suggested in [252] for H, H_2 and He targets for projectiles with $q > 5$ and energies $E = 1\,eV/u$–$10\,MeV/u$.

The molecular-to-atomic-target ratio of the capture cross sections $\sigma(H_2)/\sigma(H)$ is relevant in many applications in plasma and accelerator physics, for example in predicting the optimal conditions (type of gaseous target and its pressure) for stripping low charges heavy ions into highly charged ions with high efficiency. Recently, measurements of non-equilibrium and equilibrium charge state fractions were carried out at the UNILAC accelerator at GSI, Darmstadt, Germany for stripping of 1.4 MeV/u U, Ti and Ar ions in different gases: H_2, He, Ne, N_2, O_2, Ar, and CO_2 [65]. The results for equilibrium charge-state distributions for uranium ions with energies 1.4 and 0.74 MeV/u are given in Fig. 4.6, showing that at energy of 1.4 MeV/u the H_2 gas is the best stripper leading to creation of uranium stripped ions with the mean charge $\bar{q} \approx 29$.

In [254], an investigation of an *atomic* hydrogen gas as a stripper for 1.4 MeV/u-U^{4+} ions with final charge states $q = 4$–40 was considered using the following relations between EL and EC cross section for H_2 and H targets:

$$\sigma_{EL}(H_2)/\sigma_{EL}(H) = 2.0, \qquad (4.27)$$

$$\sigma_{EC}(H_2)/\sigma_{EC}(H) = 3.8, \qquad (4.28)$$

where (4.27) for electron loss cross sections follows the Bragg's rule, and (4.28) for electron capture from (4.25). Experimental and theoretical charge-changing cross section used for calculation in [254] are given in [255].

Figure 4.7 shows calculated equilibrium fractions for H_2 and H targets at low (10^{-4} mbar) and high (100 mbar) gas pressures eye-guided by solid and dashed curves, respectively. As it is seen, the lowest possible mean-charge state $\bar{q} \approx 26$

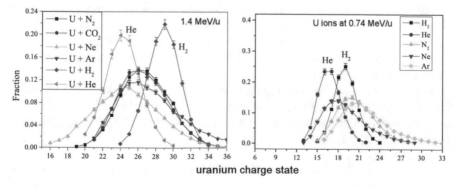

Fig. 4.6 Experimental charge-state distributions for U ions on gaseous targets at 1.4 MeV/u (left) and 0.74 MeV/u (right) energies. From [65]

Fig. 4.7 Calculated
equilibrium charge-state
fractions of 1.4 MeV/u-U^{4+}
ions, $q = 4$–40, in H_2 and H
stripper gases at low $P =
10^{-4}$ and high 100 mbar gas
pressures. From [254]

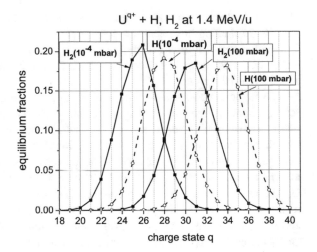

can be obtained in H_2 gas, and the highest $\bar{q} \approx 34$ in H gas as a stripper. The different ranges of q for the charge-state distributions in H_2 and H targets are due to the different relations for EL and EC cross sections for H_2 and H targets, (4.27) and (4.28). We note that if the coefficient 2.0 would be used for both EL and EC cross sections instead of the value 3.8 for electron capture the fraction distributions for H_2 and H targets would practically coincide and give the same equilibrium charges $\bar{q}$. It would constitute an interesting check of the theory if charge-state distributions for atomic and molecular hydrogen stripper would obtained experimentally.

4.5 Multiple-Electron Capture in Collisions of Heavy Ions with Gas/Solid Targets

As for multiple-electron capture (MEC) processes, involving fast heavy highly charged ions, it was found experimentally that the total capture cross section increases with increasing projectile charge, but the general cross-section dependencies on energy and atomic structures of the target and projectile have not yet been investigated experimentally in detail.

Experimental data on multiple-electron capture at very low energies ($E = 0.01$ eV/u–10 keV/u) are presented in [256–266]. For heavy low-charged ions, 268 cross sections in collisions with *noble* gases and molecules are reported in [249, 256] for the following reactions:

$$X^{q+} + A \rightarrow X^{(q-k)+} + A^{k+}, \quad 1 \leq k \leq 4, \quad q \leq 7, \tag{4.29}$$

$$X = \text{Ne, Ar, Kr, Xe}, \tag{4.30}$$

$$A = \text{He, Ne, Ar, Kr, Xe, } H_2, N_2, O_2, CH_4, CO_2. \tag{4.31}$$

These data demonstrate a semi-constant behavior of the capture cross sections at low energies. Experimental cross-section data can be fitted by a simple formula within 35% accuracy as [249]:

$$\sigma_{q,q-k} = 10^{-12} \, [\text{cm}^2] \cdot C(k) \cdot q^{A(k)} \cdot (I_T/\text{eV})^{-B(k)}, \tag{4.32}$$

$$1 \le k \le 4, \quad q \le 7, \quad E < 25 \, \text{keV/u}, \tag{4.33}$$

where I_T is the ionization potential of the target atom, and C, A, and B are the fitting parameters given in Table 4.1.

Experimental data on MEC of Xe^{q+} ions, $15 \le q \le 43$, colliding with He, Ar, and Xe atoms at slow collision velocities $v = 0.1$–0.2 a.u. are presented in [261], and these data are approximated by a semiempirical formula in the paper [262]. For slow ion collisions, $v \ll 1$ a.u., with atoms and molecules, experimental multiple-electron capture cross sections are also given in [263–267]. It is also worth noting the results of experimental [268–270] and theoretical [268, 271] studies of single- and multiple-electron capture cross sections involving *fullerenes* (hexagonal carbon rings); these data are of a particular interest for electron capture in ion collisions with complex targets (Fig. 4.8).

Table 4.1 Fitting parameters for experimental k-fold electron capture cross sections (4.32) at small ion velocities $v \ll 1$ a.u. From [249]

k	$C(k)$	$A(k)$	$B(k)$
1.	1.43 ± 0.76	1.17 ± 0.09	2.76 ± 0.19
2.	1.08 ± 0.95	0.71 ± 0.14	2.80 ± 0.32
3.	$(5.50 \pm 5.8) \times 10^{-2}$	2.10 ± 0.24	2.89 ± 0.39
4.	$(3.57 \pm 8.9) \times 10^{-4}$	4.20 ± 0.79	3.03 ± 0.86

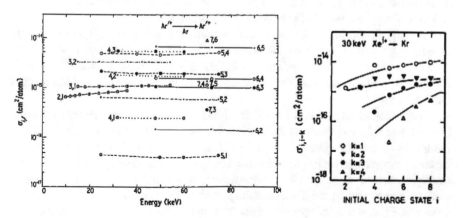

Fig. 4.8 Left: Experimental cross sections for electron capture of highly charged Ar ions in Ar target as a function of ion energy. Right: Electron capture cross sections for 30 keV Xe^{i+} ions incident on Kr. Symbols—experimental data. The solid lines correspond to (4.32). From [249, 256]

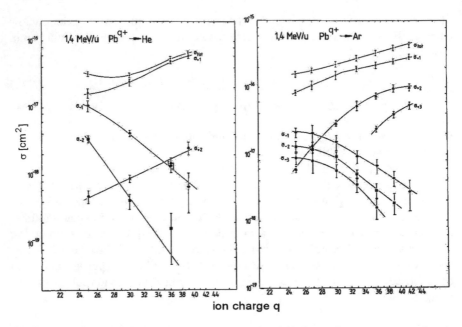

Fig. 4.9 Experimental k-fold electron capture σ_{+k} and m-fold electron loss σ_{-m} cross sections in collisions of 1.4 MeV/u-Pb^{q+} ions with He and Ar targets as a function of the ion charge q. σ_{tot} is a sum of all capture and loss cross sections. From [95]

At higher energies, $E = 1$–10 MeV/u, experimental MEC cross sections are mainly obtained for heavy Xe, Pb, and U ions colliding with gaseous targets (see [95, 96, 102, 103]. As an example, Fig. 4.9 shows experimental electron-capture and loss cross sections in collisions of 1.4 MeV/u-Pb^{q+} ions with He and Ar targets [95] as a function of ion charge q. It is seen that the contribution of multiple-electron cross sections is large, especially for heavy target atoms (Ar).

The importance of MEC processes is also demonstrated in Table 4.2 where experimental partial and total cross sections of U^{q+} ions colliding with Ar at 3.5 MeV/u is shown. It is seen that for highly charged uranium ions U^{53+} a contribution of multiple-electron capture processes to the total cross sections reaches close to 40%.

Theoretical investigations of MEC are rather limited and refer mainly to low and intermediate collision energies. Here one has to mention the classical Bohr-Lindhard model [180], developed later in papers [272, 273]; this model is primarily used for capture processes involving few-electron projectiles. Numerical calculations of the multiple-electron capture cross sections have been generally performed for double-electron capture using the close-coupling method [274–276], the quasi-molecular model [278], and the independent-particle model (IPM) [279]. As for theoretical models for multiple- electron capture $k > 2$ by heavy ions from atoms, one has to admit that at present they are somewhat incompletely developed.

Table 4.2 Experimental partial and total electron-capture cross sections (in 10^{-18}cm^2) of U^{q+} ions in collisions with Ar at 3.5 MeV/u

Reaction	Energy, MeV/u	σ_1	σ_2	σ_3	...	σ_{tot}	$\sum_m \sigma_{m\geq2}/\sigma_{tot}$, %
U^{28+} + Ar	3.5	12.6	–	–	...	12.6	0
U^{31+} + Ar	3.5	19.7	1.1	–	...	20.8	5.3
U^{33+} + Ar	3.5	25.0	2.0	–	...	27.3	7.5
U^{39+} + Ar	3.5	52.3	8.1	0.30	...	60.7	13.8
U^{42+} + Ar	3.5	61.6	16.1	2.0	...	79.7	22.7
U^{53+} + Ar	3.5	82.5	35.3	10.6	...	129.8	36.4

Relative contributions $\sum_m \sigma_{m\geq2}/\sigma_{tot}$ of multiple-electron capture processes to the total cross sections are also given. From [103]

Fig. 4.10 Scaled double-electron capture cross sections $\sigma_2(EC)/q^{0.5}$ of positive ions with charge state $q \geq 2$ on He atoms as a function of scaled ion energy E/q. Solid circles—experimental data, solid curve—fitting formula (4.36). From [197]

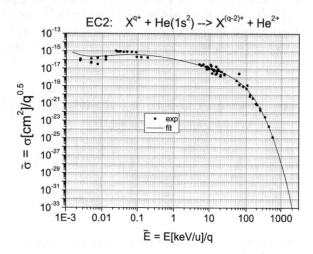

Finally, we have to mention about double-electron capture on He atoms. The corresponding cross sections $\sigma_2(EC)$ are measured for many projectiles over a wide energy range (see [197] for references) for reactions:

$$X^{q+} + He(1s^2) \rightarrow X^{(q-2)+} + He^{2+}, \quad q \geq 2. \tag{4.34}$$

In [197] experimental data for $\sigma_2(EC)$ values in He target are analyzed and presented in the scaled coordinates (see Fig. 4.10):

$$\tilde{\sigma} = \sigma_2(EC)[\text{cm}^2]/q^{0.5}, \quad \tilde{E} = E[\text{keV/u}]/q, \tag{4.35}$$

where q and E denote the incident ion charge and energy, respectively. The solid line in the figure represents the polynomial fit of the scaled experimental data by the formula:

$$\log_{10} \tilde{\sigma} = \sum_{k=0}^{5} A_k \, (\log_{10} \tilde{E})^k, \tag{4.36}$$

whose fitting parameters are: $A_0 = -15.92026$, $A_1 = -0.79322$, $A_2 = -0.21674$, $A_3 = +0.06236$, $A_4 = -0.04224$, $A_5 = -0.02414$. As seen from Fig. 4.10, the accuracy of the semiempirical formula (4.36) is a factor of 2–3.

Semiempirical formula (4.36) for He target can be used to estimate the double-electron capture cross sections $\sigma_2(EC)$ for arbitrary projectile ions except for He^{2+} projectile: because of the resonance character of this reaction, the corresponding cross section $\sigma_2(EC)$ is about one order of magnitude higher than that given by the semiempirical formula (4.36). Resonant high-energy double-electron capture processes are considered in various papers (see, e.g., [279, 280]), and double capture processes of slow ions colliding with excited helium atoms He*$(1s2\ell)$ are considered in [281]. Again, the semiempirical formula (4.36) and Fig. 4.10 for double-electron capture cross sections can be used, e.g., for calculation of the mean charge states of heavy and super-heavy ions passing through He-filled separator (see Sect. 9.3).

4.6   Electron Capture at Relativistic Energies. Radiative Electron Capture (REC)

As has been considered before (Sect. 1.4), as the ion energy increases, the radiative electron capture REC processes

$$X^{q+} + A \rightarrow X^{(q-1)+} + A^+ + \hbar\omega_{REC}, \tag{4.37}$$

begin to contribute to the total capture cross section

$$\sigma_{tot} = \sigma_{NRC} + \sigma_{REC}. \tag{4.38}$$

where $\hbar\omega_{REC}$ denotes a photon emitted in the REC reaction, σ_{NRC} and σ_{REC} denote cross sections of non-radiative capture (NRC) and radiative-electron capture, respectively. REC is similar to radiative recombination process (RR) but is accompanied by capture of bound electrons of the target atom. REC processes are considered in detail in [40].

NRC processes more strongly depend on the atomic number than REC and decease more rapidly with the ion energy:

$$\sigma_{NRC} \sim q^5 Z^5/E^{5.5}, \quad \sigma_{REC} \sim q^5 Z/E, \quad v^2 >> I_T, \tag{4.39}$$

where Z and I_T denote the atomic number and ionization potential of the target atom.

At high relativistic energies

$$\gamma = \frac{1}{\sqrt{1-(v/c)^2}} > 10, \quad \beta > 0.99, \tag{4.40}$$

electron capture of ions on atoms is accompanied by creation of electron-positron pairs (see [40, 282]), but these processes are not considered here.

First measurements of the REC cross sections were performed in [283–285] at BEVALAC accelerator in Lawrence Berkeley National Laboratory (USA) and later, with development of powerful heavy-ion accelerators and other devices such as SUPER-EBIT in Berkeley and UNILAC in Darmstadt, Germany (see [40]).

Calculations of NRC cross sections at relativistic energies are performed in the eikonal approximation [244, 286, 287], and of REC cross sections in impulse approximation [40], density-matrix approach [282] and relativistic approximation [288, 289]. Results of calculated NRC cross sections at relativistic energies are presented in [290] and RR and REC cross sections in [291], respectively.

In Fig. 4.11, experimental and theoretical total electron capture cross sections for H-like Au, Bi and U projectiles in C, Ni, Cu and Au targets are presented in

Fig. 4.11 Experimental and calculated electron-capture cross-sections σ/q^5 for H-like gold, bismuth and uranium projectiles in C, Ni, Cu and Au targets at different ion energies [117]. The experimental data (symbols) are compared with theoretical predictions REC (dashed lines), NRC (dotted lines) [289]. The full lines represent the sum of both capture processes. From [117]

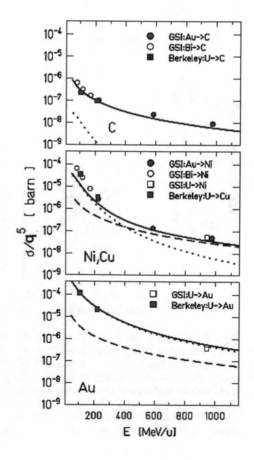

the energy range from 80 to 1000 MeV/u. The theoretical cross sections for REC, NRC [289] and the sum of both are compared with the experiment. For the C target the REC process dominates, and for Ni and Cu targets the contribution of REC is smaller up to 300 MeV/u and for Au the REC is one order of magnitude smaller than the NRC in the whole energy range. The NRC cross sections in the figure are calculated using the relativistic eikonal approximation [290]. Experimentally each contribution can be distinguished by measuring the X-rays ($\hbar\omega_{REC}$) in coincidence with the corresponding charge state after electron capture.

Another examples of different contributions of NRC and REC processes to the total capture cross sections corresponding to non-relativistic and low relativistic regimes are shown in Fig. 4.12. These calculations are based on the eikonal approximation [244, 286, 287].

At energies $E < 1$ GeV/u, a non-relativistic theory by Stobbe for the REC based on the dipole approximation [292] describes experimental data quite well, but at $E > 1$ GeV/u the relativistic theory describes the data better as was shown, e.g., in [293, 294].

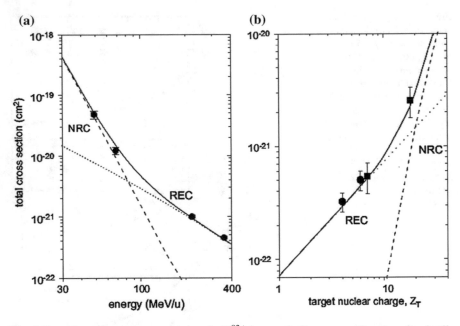

Fig. 4.12 **a** Electron-capture cross sections for U^{92+} ions on the N_2 target as a function of projectile energy. The dashed line represents the result of the eikonal approximation for the NRC process [244, 287]. The dotted line gives the prediction for REC within the dipole approximation. The solid line corresponds to the sum of both contributions. **b** Electron-capture cross sections for bare U^{92+} ions at 295 MeV/u colliding with gaseous targets: solid squares ($U^{92+} \rightarrow N_2$, Ar) and with solid targets: solid circles ($U^{92+} \rightarrow$ Be, C). For N_2 the cross section per atom is given. The results are compared with the theoretical cross sections for the NRC and the REC processes (dashed and dotted line, respectively). The total electron-capture cross sections are given by the solid line. From [295]

Fig. 4.13 X-ray spectrum observed at nearly 150° for $U^{92+} \rightarrow N_2$ collisions at 310 MeV/u. The data were taken in coincidence with outgoing U^{91+} ions. From [298]

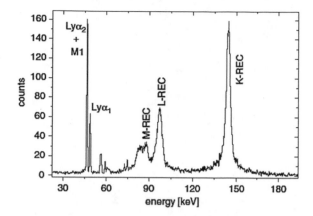

In practice, the REC cross sections are estimated by the Kramers formulae, which are applied also to radiative recombination cross sections, capturing free (not bound like in REC) electrons (see Sect. 7.1 and formulae (7.7)–(7.9) in it).

X-ray spectra of REC photons constitute important part in spectroscopy of highly charged ions, in particular, in determination of radiation polarization, photoionization of heavy ions, radiation transition probabilities, influence of QED effects and so on (see [289, 296, 297]).

An example of X-ray spectrum of REC is displayed in Fig. 4.13. The spectrum was obtained at an observation angle of 150° for 310 MeV/u $U^{92+} \rightarrow N_2$ collisions. Besides the Lyman ground-state transitions $Ly\alpha_1$ ($2p_{3/2} \rightarrow 1s_{1/2}$), $Ly\alpha_2$ ($2p_{1/2} \rightarrow 1s_{1/2}$) and M_1 ($2s_{1/2} \rightarrow 1s_{1/2}$), the important features observed in the spectrum are radiative electron capture into the ground and excited projectile states. The widths of these lines reflect the Compton profile of the bound target electrons (see, e.g., [295]).

Chapter 5
Charge Exchange in Slow Ion-Atom Collisions. Adiabatic Approach

Abstract This chapter is devoted to consideration of electron-capture processes at very *low* collision velocities $v \ll 1$ a.u. At present time, these processes are of a high importance because of two main reasons: first, they constitute the dominant mechanisms in a low-temperature plasma for creating the impurity ions in excited states, radiation short-wavelength spectra of which are used for plasma diagnostics. Second, at low-energy collisions, electron-capture cross sections, occurring in collisions with hydrogen isotopes (H, D, and T), are strongly influenced by the so-called *isotope effect* which changes the cross sections values by orders of magnitude. This influence is important for estimating, e.g., the capture cross sections of tungsten atoms and ions, colliding with neutral plasma atoms, because tungsten is adopted now as the most perspective element for making walls and diverter in plasma devices aiming at magnetic plasma confinement.

5.1 Introductory Remarks

In penetration of projectiles through low-temperature plasmas (near-wall plasma and plasma in a divertor of tokamaks and stellarators), charge exchange (electron capture) constitutes the dominant process in the population of excited states of heavy ions in plasma and, therefore, plays an important role in ion charge-state distribution, radiative cooling, and the transport of particles. The *adiabatic approach* is a powerful tool for the calculation of the charge exchange cross sections in slow ion-atom collisions. Comparing to close-coupling methods, adiabatic approach has a strong advantage: it is based on the adiabatic theory of transitions in slow collisions [28] but not on the numerical methods. In adiabatic theory, there are no assumptions on the specific form of the electronic Hamiltonian, and only the smallness of the relative nuclear velocity is used. It leads to analytical expressions for the transition probabilities, which results in a deeper understanding of the nature of nonadiabatic transitions.

The *isotope effect*, found in slow collisions of α-particles with hydrogen isotopes [299], is considered here and it is shown that the strong isotope effect exists also for heavier projectiles. The results obtained, apart from their general interest for collision

© Springer International Publishing AG 2018
I. Tolstikhina et al., *Basic Atomic Interactions of Accelerated Heavy Ions in Matter*, Springer Series on Atomic, Optical, and Plasma Physics 98, https://doi.org/10.1007/978-3-319-74992-1_5

physics, have important implications for diagnostics and simulation of elementary processes in fusion edge plasmas.

At present, the charge exchange process with tungsten projectiles is of great interest, since W is selected as a key material of the plasma facing components in the ITER tokamak, where tritium plasma is planned to be used [300, 301]. Here, the charge exchange cross sections for W ions calculated in the adiabatic approach are presented for the low-energy collisions [302].

A part of this chapter is devoted to the role of the *electron-nuclear* interaction in slow collisions. This interaction affects the nuclear trajectories which, in turn, affects theoretical results for the cross sections of various collision processes. The results are especially sensitive to the details of the internuclear dynamics in the presence of a strong isotope effect on the cross sections for the charge transfer in low-energy collisions with H, D, and T [299]. A potential, governing the ion-atom collision dynamics, is introduced in the frame of the adiabatic approach [303]. The potential accounts for the interaction between the electron and nuclei by adding to the Coulomb potential the potential of the initial electronic state obtained by solving the three-body Coulomb problem. It is shown that the use of the 'adiabatic' trajectory instead of the Coulomb one in the charge exchange calculations within the adiabatic approach improves the agreement of the results with ab initio calculations.

Also, the resonance charge exchange in slow (the center-of-mass energy $E \lesssim 1$ a.u.) proton-hydrogen collisions is discussed. Due to its large cross section, this process has significant influence on the energy and momentum transfer in low temperature plasmas in the divertor region of magnetic confinement fusion devices. Recent experiments on cold plasmas diagnostics have shown a necessity to include in simulations the resonance charge exchange between excited states [304, 305].

This chapter also describes an application of the adiabatic approach in the interpretation of the experimental results.

5.2   Adiabatic Approach

In the theory of atomic collisions, the adiabatic approximation is used to describe electronic transitions when the collision velocity is small and the nuclear motion can be treated classically. Transitions between electronic states of the colliding atoms are described in a classical on the nuclei motion approach by the time-dependent Schrödinger equation

$$\iota \frac{\partial \psi(\mathbf{r}, t)}{\partial t} = H(\mathbf{R}) \psi(\mathbf{r}, t), \tag{5.1}$$

where $\mathbf{r}$ is a set of electronic coordinates, $H(\mathbf{R})$ is an electronic Hamiltonian of diatomic quasi-molecule, $R = \mathbf{R}(vt)$ is an internuclear distance and v is a relative nuclear velocity. In the most common form the adiabatic approximation is an asymptotic expansion of the solution of equation (5.1) in the small parameter v. In this approximation, the electron wave function is sought in the form of expansion

$$\psi(\mathbf{r}, t) = \sum_p g_p(t)\, \varphi_p(\mathbf{r}, R)\, \exp\left(-\iota \int^t E_p(R(\upsilon t'))dt'\right) \tag{5.2}$$

in eigenfunctions of instant electronic Hamiltonian

$$H(R)\, \varphi_p(\mathbf{r}, R) = E_p(R)\, \varphi_p(\mathbf{r}, R) \tag{5.3}$$

which depend on R as on a parameter included in the Hamiltonian. The eigenvalues $E_p(R)$ are called the *molecular potential* curves. Adiabatic approximation reduces to the calculation of the principal terms of the asymptotic of the expansion coefficients $g_p(t)$ when $\upsilon \to 0$.

In the adiabatic representation, the *boundary conditions* are formulated as follows: when $R \to \infty$ the adiabatic energies $E_p(R)$ tend to energy levels of isolated fixed atoms, and $\varphi_p(\mathbf{r}, R)$—to the corresponding atomic wave functions $\varphi_p^a(\mathbf{r}, R)$. Therefore, if the effect of momentum transfer is ignored, the population of atomic states $\varphi_p^a(\mathbf{r}, R)$ before and after collision coincides with $g_p(t = \mp\infty)$, and the probability of a transition from the initial atomic state $\varphi_q^a(\mathbf{r}, R)$ to the final state $\varphi_p^a(\mathbf{r}, R)$ is

$$P_{pq} = \lim_{t \to \infty} |g_p(t)|^2, \quad \lim_{t \to -\infty} g_p(t) = \delta_{pq}. \tag{5.4}$$

The probability of the transition depends on the impact parameter ρ which defines the nuclear trajectory $\mathbf{R}(t)$. Integration over ρ gives the cross section of the *inelastic transition*

$$\sigma_{pq} = 2\pi \int_0^\infty P_{pq}(\rho)\rho d\rho, \tag{5.5}$$

which is the most important characteristic of the collision process.

The range of collision energies where the adiabatic approximation is applicable essentially depends on the process considered. From below, it is limited by the condition of the classical description of the nuclear motion, and the upper limit is a condition for the applicability of the asymptotic expansion in small υ.

To illustrate the adiabatic approach for calculating the charge exchange cross sections in slow collisions, the following reaction is considered

$$Li^{3+} + H(D, T)(1s) \to Li^{2+}(nl) + H(D, T)^+. \tag{5.6}$$

Figure 5.1 shows electronic energies for the states that are related to this reaction. These energies are the eigenvalues of the two-center Coulomb problem [307]. The problem is separable in the prolate spheroidal coordinates and is solved for the complex internuclear distance R. To denote the quasi-molecular states in the figure the spherical quantum numbers of the united atom limit are used. In the adiabatic theory, a rapid change of the electronic wave function induces the charge exchange transitions. It happens at the internuclear distances where *non-adiabatic coupling* has its maximum.

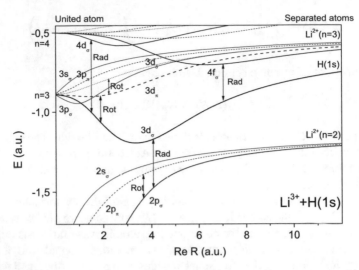

Fig. 5.1 Potential curves $E(R)$ for the charge exchange process $Li^{3+} + H(1s) \to Li^{2+}(nl) + H^+$ as a function of the internuclear distance R (in a.u.). From [306]

Transitions between two states caused by the radial coupling occur in the *crossings* (branch points) at complex internuclear separation R where electronic energies of the states are equal. Hidden crossings, which arise when the full-dimensional classical trajectory of the electron collapses into an unstable periodic orbit, are invisible on the plot of the adiabatic potential curves at the real value of the adiabatic parameter R. It requires the direct calculation in the complex R-plane. In Fig. 5.1 the 'radial' transitions are indicated by arrows (Rad) locating at real R values of the correspond- ing hidden crossings and connecting initial ($H(1s)$ in separated atom limit) and final states of the electron.

Here, the resonance channel $H(1s) + Li^{3+} \to H^+ + Li^{2+}(n = 3)$, which is responsible for the isotope effect, is considered. There are three 'radial' transitions: at the internuclear distance 1.5, 4, and 7 a.u. Since the hidden crossing at $Re(R)$ = 1.5 a.u. belongs to S-series (the so-called *super promotion* to continuum), which pairs $E_{nlm}(R)$ and $E_{n+1lm}(R)$ electronic states (n, l, m are the principle, orbital and magnetic quantum numbers) and are responsible for the ionization process in the adiabatic approximation, the corresponding transition does not contribute to the charge exchange process. Two other transitions change the charge state, and the one, between $3d_\sigma$ and $4f_\sigma$ states, is the main radial transition for the resonance charge exchange channel.

Rotational coupling is associated with internuclear axis rotation in close collisions and induces transitions between $E_{nlm}(R)$ and $E_{nlm\pm1}(R)$ electronic states. These states are degenerate in the united atom limit. An exact crossing of potential curves of these states occurs at complex R values ($Re(R) = 0$). In Fig. 5.1 'rotational' transitions are shown by arrows (Rot) located at arbitrary values of R.

5.3   Code ARSENY

The calculation of the charge exchange cross sections in slow collisions is performed using the ARSENY code [308], based on the method of hidden crossings. In the adiabatic approximation, radial inelastic transitions take place in the regions of the closest approach of potential curves and are decomposed into a sequence of individual two-level transitions via hidden crossings. First, the two-center Coulomb problem is solved in the code in the complex R-plane to calculate the adiabatic potential curves. Then all branching points are found and the corresponding *Stückelberg parameter* is calculated:

$$\Delta_{pq} = \left| \mathrm{Im} \int_{\mathrm{Re}R_c}^{R_c} \left[E_p(R) - E_q(R) \right] \frac{dR}{\upsilon(R, \rho)} \right|, \qquad (5.7)$$

where p and q is the set of quantum numbers, R_c is a complex branch point, E_p and E_q are the energies of the final and initial electron states, respectively, $\upsilon(R, \rho)$ is the radial internuclear velocity and ρ is the impact parameter.

The probability P_{pq} is calculated for the entire set of non-adiabatic transitions as a function of the impact parameter:

$$P_{pq} = \exp(-2\Delta_{pq}). \qquad (5.8)$$

The S-matrix calculated in the code is a product of elementary S-matrices corresponding to the individual transitions induced by the separated branch points. Starting with the initial S-matrix $S_{ij}^{(n)} = \delta_{ij}$, the n-th individual transition between p and q states in a brunch point at internuclear distance R_n changes the S matrix according to

$$S_{ip}^{(n)} = S_{ip}^{(n-1)}(1 - P_{pq}) + S_{iq}^{(n-1)} P_{pq}, \qquad (5.9)$$

$$S_{iq}^{(n)} = S_{iq}^{(n-1)}(1 - P_{pq}) + S_{ip}^{(n-1)} P_{pq}. \qquad (5.10)$$

A complete set of the cross sections of transitions between arbitrary initial and final states are found by integration of the S-matrix over the impact parameter:

$$\sigma_{qq} = 2\pi \int_0^\infty \left| 1 - S_{qq} \right|^2 \rho d\rho, \qquad (5.11)$$

for *elastic* scattering and

$$\sigma_{pq} = 2\pi \int_0^\infty \left| S_{pq} \right|^2 \rho d\rho \qquad (5.12)$$

for *inelastic* transition, where S_{pq} are the S-matrix elements.

The amplitudes of m-changing transitions induced by rotational interaction are found by numerical solution of the time-dependent Schrödinger equation for the Coulomb trajectory in the united atom approximation. In the present approach,

Fig. 5.2 A Coulomb
trajectory for close collisions

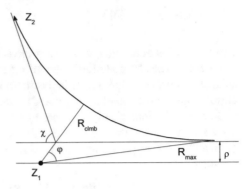

rotational interaction is considered when the internuclear separation is less than R_{max} (Fig. 5.2) given by

$$R_{max} = \frac{(l + \frac{1}{2})^2}{Z_1 + Z_2},$$ (5.13)

where l is the electron orbital momentum, Z_1 and Z_2 are the nuclear charges. This condition sets the boundary of the united atom region. The scattering angle is expressed as

$$\chi = 2 \arctan \left(\frac{Z_1 Z_2}{\mu \rho v^2} \right),$$ (5.14)

and the angle of the internuclear axis rotation as

$$\varphi = \arccos \left(\sin \frac{\chi}{2} + \frac{\rho}{R_{max}} \cos \frac{\chi}{2} \right).$$ (5.15)

For the Coulomb trajectory one has:

$$R = \frac{\rho \cos \frac{\chi}{2}}{\cos \varphi - \sin \frac{\chi}{2}}.$$ (5.16)

The internuclear separation corresponding to the closest approach is

$$R_{clmb} = \frac{\rho \cos \frac{\chi}{2}}{1 - \sin \frac{\chi}{2}}.$$ (5.17)

The dependence of the scattering angle on the reduced mass μ causes the difference in trajectories of the heavy particles in the reactions with H, D, and T and, consequently, leads to a difference in the corresponding cross sections. The amplitude of the rotational transition is the solution of the time-dependent Schrödinger equation ($\rho < R_{max}$ and $R_{max} > R_{clmb}$):

$$i\dot{a}_m - E_m a_m + i \sum_{m'=-l}^{l} \left\langle \phi_{nlm} \left| \frac{\partial}{\partial t} \right| \phi_{nlm'} \right\rangle a_{m'} = 0, \tag{5.18}$$

$$E_m = 3\gamma m^2 R^2, \quad \gamma = \frac{Z_1 Z_2 (Z_1 + Z_2)^2}{n^3 l(l+1)(2l-1)(2l+1)(2l+3)}, \tag{5.19}$$

$$\sum_{m'=-l}^{l} \left\langle \phi_{nlm} \left| \frac{\partial}{\partial t} \right| \phi_{nlm'} \right\rangle = \tag{5.20}$$

$$\frac{1}{2} \frac{\partial \varphi}{\partial t} \left[\sqrt{(l+m)(l-m+1)} \delta_{m',m-1} + \sqrt{(l-m)(l+m+1)} \delta_{m',m+1} \right],$$

where ϕ_{nlm} are the adiabatic wave functions. In the ARSENY code, they are expressed in terms of the spherical functions in the fixed coordinate system using the Wigner d-functions [28].

5.4 Influence of the Isotope Effect on Charge Exchange of Light Ions in Slow Collisions

The theory of slow ion-atom collisions has a wide application in plasma physics, controlled fusion and astrophysics. Although the study of heavy particle collisions has a long history and the theory is considered to be well established [30], new results continue to appear. A considerable difference in the cross sections of charge exchange process in collisions of He^{2+} ions with H, D, T atoms (isotope effect) has been theoretically found at very low collision energies (1–500 eV/amu) [299]. This effect is caused by the rotational interaction of electronic states in close collisions when the internuclear distance R is small. This interaction couples the electronic states of the same parity with the same principal and orbital quantum numbers n, l and the magnetic quantum numbers m, differing by unity ($\Delta m = \pm 1$). The amplitude of m-changing transition due to rotational interaction depends on the trajectories of the colliding particles which, in turn, depend on their reduced masses: the heavier the isotope the larger the charge exchange cross section.

Non-adiabatic transitions due to rotational interaction between quasi-molecular states, degenerate in the united atom limit ($R = 0$), represent the transitions of non-Landau-Zener type [28] and are of interest from the theoretical point of view. The effect of the rotational coupling on the cross sections of the charge exchange in slow collisions was studied in [309] where $He^{2+} + H(1s)$ charge exchange reaction was considered. It was shown that at collision energy below 1 keV/amu transitions due to the rotational mixing in close collisions give the main contribution to the charge exchange cross section.

The strong isotope effect caused by rotational interaction was found and investigated for $He^{2+} + H, D, T$ processes [299, 310, 311] using theoretical approach known as the Electron Nuclear Dynamics (END) [312]. This processes are of considerable interest for plasma modeling in a fusion devices. In the END approach the time-dependent Schrödinger equation is solved with trajectories of heavy particles, determined by the scattering potential, which develops in accordance with the dynamics of the electrons.

The isotopic effect is manifested at collision energies for which the adiabatic theory is applicable, so the adiabatic approximation is a natural theoretical basis for studying the effect. It was applied to study the influence of the isotope effect on the charge exchange process in slow collisions of the H, D, T isotopes with Li, Be, and C ions, and in the inverse reactions [306]. These processes play a key role in the transport of particles and in the distribution of the charge states of impurity atoms in $D - T$ burning plasmas. The adiabatic approach is valid when the energy of the system $\varepsilon = \mu v^2/2$ is much higher than the energy of the electron in the initial state. This condition determines the lower limit of the collision energy for the reactions under study: $E \gtrsim 10\,\text{eV/amu}$. In [306], to calculate the probabilities and cross sections of charge exchange the ARSENY code [308] was used. The amplitudes of m-changing transitions due to rotational mixing was found by numerical solution of the time-dependent Schrödinger equation for the Coulomb trajectory in the united atom approximation.

In Fig. 5.3, the cross sections for the charge exchange process

$$He^{2+} + H(1s) \rightarrow He^+(n = 2) + H^+, \tag{5.21}$$

calculated by the ARSENY code, are compared with two other theoretical results obtained by the hyperspherical close-coupling (HSCC) method based on the solution of the three-body Coulomb problem [313] and the Electron Nuclear Dynamics (END) method [312]. The result [306] agrees well with the HSCC calculations [313] (the most precise data available now) for energies above 60 eV/amu which corresponds to the evaluation of the applicability of our method. For lower energies, where a quantum approach should be used, the cross sections calculated with [306] method are

Fig. 5.3 Charge exchange cross sections as a function of collision energy for the reaction $He^{2+} + H(1s) \rightarrow$ $He^+(n = 2) + H^+$: present calculations [306], Chien-Nan Liu et al. [313], Stolterfoht et al. [299]

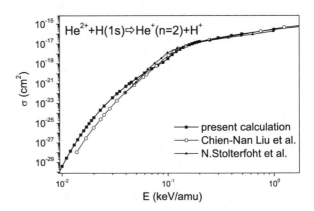

overestimated, although the dependence of the cross section on energy is described correctly. Later in this chapter, the cross sections of discussed charge exchange reactions are calculated for collision energies $E > 10\,\text{eV/amu}$. Unfortunately, there is no other data for comparison.

Figure 5.4a shows the total cross section for the reactions $Li^{3+} + H(D, T)(1s) \rightarrow Li^{2+} + H(D, T)^{+}$ calculated with the ARSENY code. The charge exchange cross section, denoted as 'w/o P^{R}' in the figure, are calculated without taking into account the rotational coupling. As can be seen from the figure, the rotational coupling starts to affect the cross section when the energy decreases below $10\,\text{keV/amu}$. This contribution increases with decreasing energy and is larger for the heavier isotope. Results of the calculation for the reaction with H target are in a good agreement with experimental data from [314]. The effect of the rotational coupling on the charge exchange cross sections diminishes with further energy decrease. At the energies below $60\,\text{eV/amu}$ rotational interaction no longer affects the cross section of charge exchange on the H target, but still completely determines the cross sections for D and T targets. It can be explained by the fact that for the collision of Li^{3+} ion with H target the internuclear separation R_{t}, which satisfies the following condition (the turning point, L is the total orbital momentum of the system) [17]:

$$\varepsilon = \frac{(L + \frac{1}{2})^2}{2\mu R_t^2} + \frac{Z_1 Z_2}{R_t}, \tag{5.22}$$

is larger than R_{max} (5.13). Consequently, for these collision energies Li^{3+} projectile does not reach the region where rotational coupling occurs.

The probabilities as functions of the impact parameter, averaged over the Stückelberg oscillations, for the resonance channel of reaction (5.6) (capture into $n = 3$ level of Li^{2+} ion) are shown in Fig. 5.4b, c for the collision energies $E_1 = 0.04\,\text{keV/amu}$ and $E_2 = 0.1\,\text{keV/amu}$, correspondingly (see in Fig. 5.4a). At collision energy E_1 the probabilities of the rotational transition $3d_\sigma - 3d_\pi$ define the electron capture probabilities for D and T targets, while for H target the probability coincides with the one calculated without rotational coupling is and close to zero. Also, as is seen from

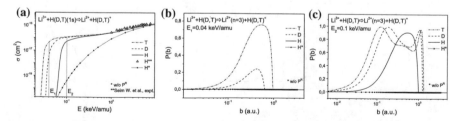

Fig. 5.4 Total charge exchange cross sections for the reaction $Li^{3+} + H(D, T)(1s)$ as a function of collision energy, E_1, E_2—see text, H^*—without otational interaction, H^{**}—experimental results from [314](a); (b), (c)—probabilities as a function of impact parameter calculated with and without taking into account the rotational interaction for the the reaction $Li^{3+} + H(D, T)(1s) \rightarrow Li^{2+}(n = 3) + H(D, T)^{+}$. From [306]

the figure, range of the impact parameters where charge exchange cross section is affected by rotational coupling is wider for collisions with T target than with D one. At the energy E_2, the projectile in collisions with H target reaches the rotational interaction region as well and all three probabilities are defined by the rotational transitions.

Total charge exchange cross sections (a) and probabilities (b, c) [306] for the reaction

$$Be^{3+}(1s) + H(D, T)(1s) \rightarrow Be^{2+}(1snl) + H(D, T)^+ \qquad (5.23)$$

are shown in Fig. 5.5. For the calculation the effective charge $Z_{eff} = 3.68$ for the final state of the electron in Be^{2+} was used. It was defined using the CDW2 code [315] based on the distorted waves approximation (CDW). In the energy range where the rotational interaction dominates the main contribution to the total cross section is given by the quasi-resonant channel with final $Be^{2+}(n = 3)$ state. For this channel, the probabilities as a function of the impact parameter are plotted in Fig. 5.5b and c for the collision energies $E_1 = 0.05$ keV/amu and $E_2 = 0.1$ keV/amu. The behavior of the cross sections and probabilities for the reaction (5.23) has the same features a for the reaction (5.6) discussed above.

In the inverse processes

$$H(D, T)^+ + C(1s^2 2s^2 2p^2) \rightarrow H(D, T)(nl) + C^+(1s^2 2s^2 2p) \qquad (5.24)$$

the initial state of the electron is $2p$ state in C atom. These processes were studied and for the calculation of charge exchange cross sections and probabilities the effective charge of the initial state of the electron $Z_{eff} = 1.86$ was used [306]. Figure 5.6 shows results of the calculation. In this reactions, the main contribution to the total cross section is given by two channels with the final states $H(D, T)(n = 1)$ and $H(D, T)(n = 2) : n = 2$ for lower and $n = 1$ for higher collision energies where the rotational interaction prevails. The presence of two channels results in a non-monotonic behavior of the charge exchange cross sections.

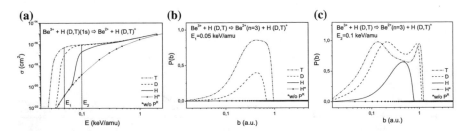

Fig. 5.5 Total charge exchange cross sections for the reaction $Be^{3+}(1s) + H(D, T)(1s) \rightarrow Be^{2+}(1snl) + H(D, T)^+$ as a function of collision energy, E_1, E_2—see text, H^*— without rotational interaction (**a**); (**b**), (**c**)—probabilities as a function of impact parameter calculated with and without taking into account the rotational interaction for the the reaction $Be^{3+}(1s) + H(D, T)(1s) \rightarrow Be^{2+}(1s3l) + H(D, T)^+$. From [306]

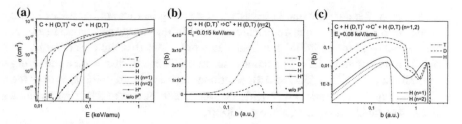

Fig. 5.6 Total charge exchange cross sections for the reaction $C(1s^2 2s^2 2p^2) + H(D, T)^+$ as a function of collision energy, E_1, E_2—see text, H^*—without rotational interaction (**a**); probabilities as a function of impact parameter calculated with and without taking into account the rotational interaction for the the reactions $C(1s^2 2s^2 2p^2) + H(D, T)^+ \rightarrow C^+(1s^2 2s^2 2p) + H(D, T)(n = 2)$ (**b**) and ($n = 1, 2$) (**c**). From [306]

Fig. 5.7 Calculated potential curves describing the charge exchange process $C(1s^2 2s^2 2p^2) + H^+ \rightarrow C^+(1s^2 2s^2 2p) + H(nl)$ as a function of internuclear distance. From [306]

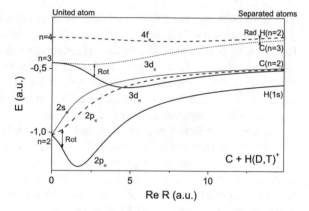

The contribution of each channel to the total cross section in collisions with H^+ projectile is shown in Fig. 5.6a. Electronic potential curves of the states which contribute to the isotope effect of the charge exchange process (5.24) are plotted in the Fig. 5.7. The states corresponding to the $2p0$ and $2p1$ initial states of the electron in C atom are the $3d_\sigma$ and $2p_\pi$ states in the united atom limit. The rotational transition $2p_\pi - 2p_\sigma$ governs the transition from the initial state $C(2p1)$ to the final state $H(D, T)(1s0)$ which corresponds to the $2p_\sigma$ in the united atom limit. Transition to the final state $H(D, T)(n = 2)$ occurs as a result of two successive transitions: rotational transition $3d_\sigma - 3d_\pi$ and radial transition $3d_\pi - 4f_\pi$ which occurs at $R = 13.7$ a.u. ($4f_\pi$ corresponds to the $H(D, T)(2p1)$ state in the separated atom limit).

In Fig. 5.6b the probabilities for the collision energy $E_1 = 0.015$ keV/amu are shown as functions of the impact parameter. There is no rotational interaction in collisions with H^+ projectile at this energy and the probabilities for the D^+ and T^+ projectiles are defined by the transitions to $n = 2$ final states. For the energy $E_2 = 0.08$ keV/amu (Fig. 5.6c), H^+ projectile also enters the region of the rotational interaction. Contributions of the $n = 1$ and $n = 2$ channels to the total cross section are shown for the H^+ projectile.

Analysis of the data obtained suggests that a strong isotopic effect arises in the presence of a resonant or quasi resonant channel in the charge exchange reaction, when the energies of the initial and final states of the electron are exactly equal or differ by a very small value. In such a case, in collisions with energies satisfying the condition $R_t < R_{max}$ (5.22, 5.13), rotational interaction predominates over the radial one, rotational transitions provide the main contribution to the cross sections and the isotopic effect occurs (see, e.g., [306]).

5.5　Charge-Changing Collisions of Tungsten and Its Ions with Neutral Atoms

The development, creation and commissioning of new powerful thermonuclear plasma installations (tokamak of the international project ITER) and the transition to the use of plasma with deuterium-tritium ignition (LHD stellarator, Japan) basically determine the fields of plasma physics that are most effectively developed in the present time. Atomic charge-changing collision processes, involving very heavy particles, are of great interest, especially collisions with Tungsten, W ($Z = 74$), while applying it in high-power nuclear-fusion studies.

Tungsten is the most promising material for manufacturing surfaces that are in contact with the plasma in high power fusion plasma devices. It was chosen as one of the key components in the plasma facing component material in ITER tokamak [316] due to its unique characteristics [300]: good thermo-mechanical properties, high melting point, low retention of tritium, and low sputtering rates. These features minimize the influx of W particles into the central high-temperature plasmas, which avoids plasma instabilities, such as plasma energy losses, radiation collapses and disruptions. On the other hand, due to very high atomic number ($Z = 74$), even a very tiny concentration of W (of the order of 0.01%) can destroy the high-temperature plasmas and cause serious disruptions.

Thus, the detailed study of atomic features of tungsten and its ions while colliding with plasma particles and knowledge of the cross sections of such collisions in a wide range of parameters (charge, state, collision energy) is necessary for clear and accurate understanding of the general characteristics (stability, ignition) of the main, near-wall and diverter plasmas in high-temperature fusion devices. Still limited data on the collisions of W atoms and ions is with plasma particles requires further research (see, e.g., [301]).

Paper [302] is devoted to the experimental and theoretical investigations of the charge-changing collisions of tungsten and its ions with neutral atoms. In [302] the charge exchange cross sections were measured for the following reactions:

$$W^+(6s) + H_2 \rightarrow W + H_2^+, \tag{5.25}$$

$$W^+(6s) + He(1s^2) \rightarrow W + He^+(1s) \tag{5.26}$$

and

$$W^{2+}(5d^4) + He(1s^2) \rightarrow W^+ + He^+(1s). \tag{5.27}$$

The measurements of the charge exchange cross sections are performed for the following collisional energies: 55 eV/u [reactions (5.25) and (5.26)] and 82 eV/u [reaction (5.27)]. These reactions, taking place in detached plasmas of fusion devices with tungsten as a PFC material, have a strong influence on plasma dynamics.

The experimental set-up for the measurements of the charge exchange cross section in reactions (5.25)–(5.27) is described in [317, 318]. In the [302], a 99.95% purity tungsten wire is bombarded with a pump beam of CO^+ ions with the energy 0.9 MeV. To perform a mass/charge analysis of the extracted ions a Wien filter is used which acted as a velocity filter. Other undesired ions of the same velocity as the W ions, which could also pass through the filter, are removed with a neutral particle rejector (NPR), composed of 4 sets of electric fields to make ions of different masses (with the same velocity) follow different trajectories. In the NPR, the ion beam trajectory is shifted aside by the upstream two electric fields according to ion mass. Only W^+ ions pass through a movable slit before restoring the straight path with the electric fields downstream. Figure 5.8 shows an example of the q/M spectrum (q is the charge and M is the ion mass), obtained by scanning the Wien filter electric field with a fixed magnetic field. The figure illustrates the clear separation of the W^+ and W^{2+} ion peaks from WO^+ and WO_2^+ peaks (the horizontal axis is proportional to the square root of q/M value). The calculation of the charge exchange cross sections are performed in adiabatic approximation using the ARSENY code [308] based on the hidden crossing method. The calculated data are compared with experimental charge exchange cross sections and results of their comparison are presented in Fig. 5.9 for the He target [reactions (5.26) and (5.27)] and in Fig. 5.10 for the H target [reaction (5.25)]. Experimental value of the charge exchange cross section in collisions of W^+ ions with He target for the energy 55 eV/u is $(3.2 \pm 0.2) \times 10^{-18}$ cm^2 and for W^{2+} ion

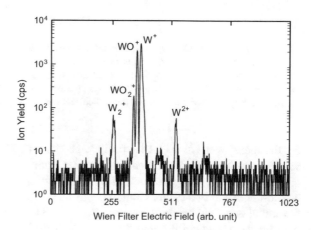

Fig. 5.8 An example of the q/M spectrum, obtained by scanning the electric field of the Wien filter. From [317]

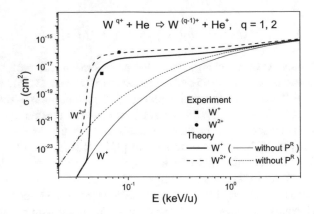

Fig. 5.9 Total charge exchange cross sections for W^+ and W^{2+} ions colliding with He atom as a function of ion energy: solid square—$W^+ + He$, experiment; solid circle— $W^{2+} + He$, experiment; thick solid line—$W^+ + He$, theory; thin solid line—$W^+ + He$, theory (without taking into account the rotational interaction P^R); thick dashed line— $W^{2+} + He$, theory; thin dashed line—$W^{2+} + He$, theory (without taking into account the rotational interaction P^R). All experimental and theoretical data are from [302], see text

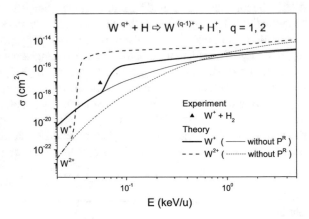

Fig. 5.10 Total charge exchange cross sections for W^+ and W^{2+} ions colliding with H and H_2 targets: solid triangle—$W^+ + H_2$, experiment; thick solid line—$W^+ + H$, theory; thin solid line— $W^+ + H$, theory (without taking into account the rotational interaction P^R); thick dashed line— $W^{2+} + H$, theory; thin dashed line—$W^{2+} + H$, theory (without taking into account the rotational interaction P^R). All experimental and theoretical data are from [302], see text

for the energy 82 eV/u is $(1.14 \pm 0.01) \times 10^{-16}$ cm^2. The corresponding theoretical values are 9.4×10^{-18} cm^2 and 9.45×10^{-17} cm^2.

The electronic energies relevant to the collisions with He target are found as a solution of the two-center Coulomb problem [307] with the effective charges of the initial and final electronic states: $Z_{eff} = 1.34$ for the $1s$ state in He, 4.56 and 6.31 for the 6s state in W^+ and W^{2+}, respectively. The calculations assume that the target and the projectile are in the initial ground state. In the adiabatic approximation the electronic transitions are induced by the radial interaction, associated with the change of the internuclear separation, and the rotational one doe to internuclear axis orientation. Here, the charge exchange cross section are calculated with and without accounting for the rotational coupling (P^R).

As is seen from Fig. 5.9, the rotational interaction starts to affect the cross sections in reactions (5.26) and (5.27) when the energy of collisions decreases below 3 keV/u and, while the energy decreases to 50 eV/u, the contribution of transitions caused by the rotational mixing to the charge exchange cross sections increases. With further energy decrease, the effect of the rotational coupling on cross sections diminishes. The main contribution to the charge exchange cross sections measured at energies 55 and 82 eV/u is given by transitions caused by rotational coupling. It can be also noted that the experimental energy 55 eV/u for measuring the cross section in collisions with W^+ projectile is close to the boundary of the energy range where the rotational coupling affects the charge exchange cross section.

The calculations [302] are performed with the Coulomb trajectory in close collisions. In [311] the END approach [312] is used to study the charge exchange process in slow collisions of He^{2+} with H. In this approach the trajectory of heavy particles differs from the Coulomb one and it was shown that the scattering angle at low collisional energies obtained by the END differs from the scattering angle for the Coulombic case. It means that changing the trajectory can lead to a change of the energy range where the rotational mixing influences charge exchange cross sections. This may be the reason for the greater difference between the experimental and theoretical results for the W^+ projectile. A good agreement between experimental and theoretical results confirms the concept used in the adiabatic approach for description of the mechanism of transitions in slow collisions.

Theoretical results for charge exchange cross sections in collisions of W^+ and W^{2+} ions with H atom as an energy functions together with experimental value of the cross section for charge exchange in $W^+ + H_2$ reaction ($\sigma_{H_2} = (8.05 \pm 0.22) \times 10^{-18} \mathrm{cm}^2$) measured at $E = 55$ eV/u are shown in Fig. 5.10. As seen in the figure, an influence of the rotational coupling on the charge exchange cross section in collisions with W^{2+} projectile is stronger and an its energy range is wider than in collisions with W^+ projectile. Unfortunately, experimental data on collisions of W^+ and W^{2+} ions with atomic hydrogen is currently absent. To evaluate the rationality of the theoretical results, the experimental result available for the cross section in reaction of W^+ with H_2 target and the data on the ratio σ_{H_2}/σ_H of the single electron capture are considered. This ratio in collisions of highly charged ions with atomic and molecular hydrogen targets can be found in [251]. It rises from 0.7 to 4 while the collisional energy increases. Results of investigation of heavy low charged ions including W^{q+} ions ($6 \leq q \leq 15$) colliding with H and H_2 targets (the case under consideration) are described in [319–321]. It was found that the ratio of charge exchange cross sections σ_{H_2}/σ_H in collisions with low charged ions differs from that for highly charged projectiles—as the energy of collisions decreases the ratio becomes greater than 1 and continues to rise [319].

In the experiment from [322], σ_H and σ_{H_2} were measured for charge exchange in collisions of Al^{2+} ions with atomic and molecular hydrogen at 185 eV/u–1 keV/u energy range. The ratio σ_{H_2}/σ_H increases from 1.2 at $E =$ 1 keV/u to 2.5 at $E =$ 185 eV/u which confirms the conclusion made in [319]. Since theoretical method [302] reproduces experimental results for atomic targets very well (for the collisions with Al^{2+} projectiles at energy 185 eV/u experimental σ_H is $(1.97 \pm 0.31) \times 10^{-16} cm^2$ and theoretical one is $1.85 \times 10^{-16} cm^2$), theoretical value of the ratio σ_{H_2}/σ_H in collisions of the W^+ projectile at energy 55 eV/u is estimated. According to experimental (σ_{H_2}) and theoretical (σ_H) values this ratio is lower than 5.4. The energy considered, 55 eV/u, is on the boundary of the range of the influence of the rotational mixing (see Fig. 5.10). If the non-Coulombic trajectory will be used in the calculations the range of the rotational coupling influence will be wider and therefore the theoretical value of the CE cross section will increase. This will reduce the ratio σ_{H_2}/σ_H but not less than 1 ([319, 322]). The role of the trajectory in adiabatic approximation is considered later in this chapter.

5.6   Isotope Effect in Low Energy Collisions of W^+ and W^{2+} ions with H, D and T

In this section, the isotope effect is studied for reactions

$$W^+(6s) + H(D, T)(1s) \rightarrow W + H(D, T)^+ \tag{5.28}$$

and

$$W^{2+}(5d^4) + H(D, T)(1s) \rightarrow W^+ + H(D, T)^+ \tag{5.29}$$

in the framework of the adiabatic theory of transitions in slow collisions [28]. These reactions can occur in detached low-temperature plasmas of fusion machines (near-wall or diverter regions) where collision energies correspond to the energy range of the isotope effect (from 5 to 500 eV/u). In ITER experiments, the DT burning plasma is planned to use, which make the influence of the isotope effect on plasma parameters an important issue. Here, the same theoretical approach [302] is used for calculations of the charge exchange cross sections for the reactions (5.28, 5.29). Results of the calculations show a significant difference between the cross sections in collisions with different isotopes (the isotope effect).

Total charge exchange cross sections in collisions of W^+ and W^{2+} ions with H, D and T targets are calculated with (Fig. 5.11a) and without (Fig. 5.11b) taking into account the rotational coupling. If neglecting the rotational interactions, the difference between cross sections in collisions with different isotopes disappears. In the reaction with W^+ projectile (Fig. 5.11a) the isotope effect is manifested when the collision energy decreases below 200 eV/u. At the energies below 56 eV/u the cross section for the collisions with H target is not affected by the rotational mixing

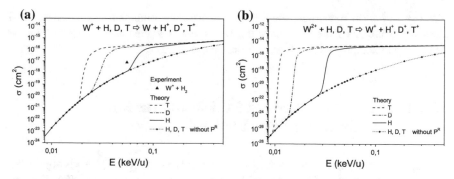

Fig. 5.11 Total charge exchange cross sections for the reactions $W^+ + H, D, T$ (a) and $W^{2+} + H, D, T$ (b): solid triangle— $W^+ + H_2$, experiment; solid line—reaction with H, theory; dashed line—reaction with T, theory; dashed dotted line— reaction with D, theory; dotted line with solid circles—reaction with H, D, T, theory (without taking into account the rotational interaction P^R). All experimental and theoretical data are from [302], see text

since the W^+ projectile does not enter the region of rotational interaction (R_{clmb} is larger than R_{max}). The same happens in collisions with D and T targets when energy decreases below 23 and 18 eV/u, respectively. The heavier the isotope the wider the energy range of the effect of rotational interaction. In collisions with W^+ projectile, the highest ratio σ_D/σ_H is about 3 and σ_T/σ_H is about 4 orders of magnitude.

As can be seen from Fig. 5.11b, not only increasing the mass of the target, but also increasing the charge of the projectile enlarges the energy range where the rotational interaction affects the charge exchange cross sections. This is due to the fact, that as the energy decreases the condition defining the rotational interaction region ($R < R_{max}$ and $R_{max} > R_{clmb}$) can be satisfied for the electronic states with larger orbital momenta. In the united atom limit, the larger orbital momentum of the electronic states correspond to the collisions with the projectile of a higher nuclear charge. Since the effective nuclear charge of W^{2+} (6.31) is higher than that of W^+ (4.56) the W^{2+} projectile can enter the rotational mixing region at lower collision energies. For the same collisional energy, the contribution of the rotational transitions to the charge exchange cross section in collisions with W^{2+} projectile is also larger because the condition ($R < R_{max}$ and $R_{max} > R_{clmb}$) is valid for a wider range of impact parameter. It results in more strong isotope effect for the W^{2+} projectile: σ_D/σ_H is about 8 and σ_T/σ_H is about 10 orders of magnitude. As far as we know, the experimental data on the charge exchange cross sections for collisions with hydrogen isotopes, which can confirm the theoretical prediction of the isotope effect [302], are still absent. The only indirect confirmation of the present calculation is the discussed above experimental result for the collision of W^+ projectile with molecular target H_2 (Fig. 5.11a).

Since an influence of the isotope effect on charge exchange cross sections is very strong in the energy range, which corresponds to the collision energies in detached plasmas of fusion machines, the data considered are indispensable in theoretical studies and numerical simulations of the $D - T$ burning plasmas behavior, and must be taken into account when planning future experiments.

5.7 Influence of the Electron-Nuclear Interaction on Charge Exchange Cross Sections

The classical treatment of the nuclei motion in slow ion-atom collisions is applicable if the nuclear interaction is described by some certain effective potential. This interaction determines the internuclear trajectory which depends on the masses of the colliding nuclei. The dependance of the trajectory on the nuclei masses results in the isotope effect. In the END approach [312] used in [299, 310, 311], the trajectories are controlled by the scattering potential, which develops in accordance with the dynamics of the electrons. At certain collision parameters, this approach produces counterintuitive result for the internuclear motion: in the reaction

$$He^{2+} + A(1s) \rightarrow He^{+}(n = 2) + A^{+}, \quad A = H, D, T, \tag{5.30}$$

the scattering angles are *negative* [311]. This result cannot be explained by any purely repulsive interaction. The adiabatic approximation uses the Coulomb trajectory of the internuclear motion with a different dependence on the collision parameters. Obviously, the presence of a strong isotopic effect leads to the dependence of the theoretical results on the trajectory, and, consequently, on the features of the internuclear interaction.

The influence of the electron-nuclei interaction on the internuclear trajectory, and therefore on the observable charge exchange cross sections in slow ion-atom collisions, was investigated in [303]. The conclusion made in the work is very clear: to describe the internuclear motion the *Born-Oppenheimer* (BO) potential corresponding to the entrance channel of the reaction should be used. It was shown that the use of BO potential, which effectively accounts for the electron-nuclei interaction, allows to explain qualitatively and reproduce quantitatively the negative scattering angles reported in [311]. This demonstrates that the BO approximation can be used to reproduce the accurate ab initio calculation of the internuclear trajectory, so significant for the END approach used in [299, 310, 311].

System under consideration consists of an electron and two nuclei with charges Z_i and masses M_i, $i = 1, 2$. It is assumed that the electron is initially bound to the second nucleus in the nondegenerate ground state with energy $E_0 = -Z_2^2/2$ in order to avoid complexities caused by the Coulomb degeneracy [323]. Let us consider a collision of the first nucleus (projectile) with the bound pair (target) with initial relative velocity of the nuclei v. The energy of the internuclear motion in the center-of-mass frame is $E = \mu v^2/2$, where $\mu = M_1 M_2/(M_1 + M_2)$ denotes the reduced mass. The interaction between two bare nuclei is described by the Coulomb potential

$$V_C(R) = \frac{Z_1 Z_2}{R}, \tag{5.31}$$

where R is the internuclear distance. The internuclear interaction in the BO approximation is described by

$$V_{BO}(R) = V_C(R) + E(R) - E_0, \tag{5.32}$$

where $E(R)$ is an eigenvalue of the two-center Coulomb problem [307], solved for the present system, and corresponds to the energy of the electron in the field of the nuclei fixed in space at an internuclear distance R. According to the *adiabatic theorem*, electronic transitions in slow collisions are suppressed. This leads to an adiabatic evolution of the electronic state, which defines the potential (5.32). It means that the electronic state is the two-center Coulomb problem eigenstate coincided with the initial state for $R \to \infty$, therefore $E(R \to \infty) = E_0$. In (5.32), the electronic terms account for the electron-nuclei interaction. Figure 5.12 shows the different terms in (5.32) for the system with $Z_1 = 2$, $Z_2 = 1$, and $E_0 = -0.5$. It can be noted that the Coulomb potential is repulsive $(dV_C(R)/dR < 0)$ for all internuclear separations R, while the BO potential has a shallow minimum at $R \approx 3.9$ and is attractive $(dV_{BO}(R)/dR > 0)$ for larger R.

The trajectories defined by the potentials (5.31) and (5.32) are different. To find out which one best describes the internuclear motion, the trajectories should be compared with reliable results. Accurate ab initio results for the considered system (5.30) were reported in [311]. These results were obtained in the END approach [312] based on the solution of the time-dependent Schrödinger equation incorporating coupled electron and nuclear dynamics. In the END approach, the internuclear trajectory is self-consistently defined by the instantaneous forces, acting on the nuclei, determined by the instantaneous distribution of the electronic density. As a result, the END approach yields a nontrivial trajectory, but not a time-independent internuclear potential.

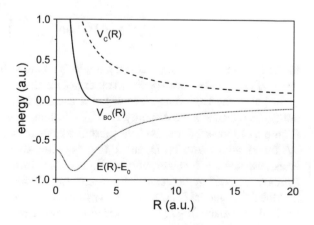

Fig. 5.12 The Coulomb (5.31) and BO (5.32) internuclear potentials for the system (5.30). From [303]

In [311], the scattering angle is considered as feature of the trajectory which characterizes the internuclear dynamics. It was shown that this angle obtained in the END approach for the system (5.30) depends on the target for the certain collision parameters. A comparison with the classical scattering angles obtained for two time-independent potentials gives the following result: the END results agree with the results for the Coulomb potential (5.31) at high collision energies ($E \gtrsim 10$ keV/amu), but at lower energies strong discrepancies occur.

The results for a screened Coulomb potential obtained by means of a screened frozen charge distribution for a target in the initial state are in closer agreement with the END results at lower energies, but only at large impact parameters. There is a qualitative difference at smaller impact parameters. Namely, it was found in [311] that the scattering angle is negative for a certain interval of impact parameters and sufficiently low energies ($E \lesssim 1$ keV/amu). Since the Coulomb and screened Coulomb potentials are purely repulsive, and the repulsive interaction leads to positive scattering angles, they cannot reproduce negative scattering angles obtained in the END approach. It can be assumed that the negative scattering angles occur due to an attractive part of the internuclear interaction.

Let us consider whether the negative scattering angles can be reproduced by the BO potential (5.32). For a given internuclear potential $V(R)$, the classical scattering angle as a function of the impact parameter ρ and collision energy E is given by [324]

$$\theta(\rho, E) = \pi - \int_{R_{min}}^{\infty} \frac{2\rho dR}{R^2 F(R)}, \qquad (5.33)$$

where

$$F(R) = \sqrt{1 - \frac{\rho^2}{R^2} - \frac{V(R)}{E}}, \qquad (5.34)$$

and R_{min} is the internuclear distance of closest approach defined by the equation $F(R_{min}) = 0$. For the Coulomb interaction potential between two bare nuclei (5.31) one obtains $\theta(\rho, E) = \theta_C(\rho E)$, where

$$\theta_C(\rho E) = 2 \arctan\left(\frac{Z_1 Z_2}{2\rho E}\right). \qquad (5.35)$$

In the Coulombic case the scattering angle is a unique function of *product* ρE, but this scaling rule does not applies to other potentials.

In [303], the BO potential (5.32) was used to calculate the scattering angle in collisions (5.30) at a fixed energy $E = 50$ eV/amu, as was also considered in [311]. Figure 5.13 shows the results of the present calculations as functions of the parameter ρE for collisions with H, D, and T targets by the solid, dashed, and dashed-dotted lines, respectively, together with the END results from [311], which are shown by symbols. A good agreement of the results for all the considered ρE values proves that the BO potential (5.32) can well reproduce the collision dynamics predicted by the END approach. It can be concluded that the negative scattering angles found in

[311] arise due to the attractive part of the BO potential caused by the inclusion of the electron-nuclei interaction. For comparison, the scattering angle for the Coulomb potential (5.31) is shown in Fig. 5.13 with the dotted line. Due to the scaling rule in the Coulombic case, there is no difference between H, D, and T targets. This line, locating rather far from the ab initio results, obviously fails to describe the negative scattering angles.

Figure 5.14 shows the Coulomb and BO trajectories calculated for the collision parameters $E = 50$ eV/u and $\rho = 4.68$ a.u. where they have different signs. The value of $\rho = 4.68$ a.u. is chosen to be close to the values of the impact parameter in the minima of the scattering angles for the BO trajectories: $\rho = 4.68$, 4.66, and 4.64 a.u. for H, D, and T, respectively (Fig. 5.13). Since these collision parameters correspond to the attractive region of the interaction, the scattering angles for the BO trajectories are negative.

Comparison of the geometrical parameters of the trajectories plotted in Fig. 5.14 is given in Table 5.1 and results in two main conclusions. First, trajectories for the different targets governed by the same potential are different, which eventually causes the isotope effect. Second, in some region of the collision parameters E and ρ, BO trajectories significantly differ from the corresponding Coulomb trajectories. It can be noted that with an increase in the target mass the distance of closet approach demonstrates opposite behaviors for the Coulomb and BO potentials. The effect of the R_{min} value on the cross sections is considered in the next section.

Let us determine whether the cross sections of charge transfer in collisions (5.30) are sensitive to the difference between the Coulomb and BO trajectories. In [303], the cross sections were calculated in the the adiabatic approach, realized in the ARSENY code. This approach is based on the asymptotical solution of the time-dependent Schrödinger equation for an electron in the three-body system for relative velocity $v \to 0$. A detailed description of the adiabatic approach is given in one of the previous

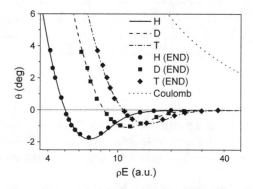

Fig. 5.13 Scattering angle for the internuclear motion in the system (5.30) at collision energy $E = 50$ eV/amu as a function of the product ρE. Solid, dashed, and dashed-dotted lines: present results for H, D, and T, respectively, calculated using the BO potential (5.32) in (5.14). Symbols: ab initio END results from [311]. Dotted line (the same for all three targets): results for the Coulomb potential (5.31) from (5.35). From [303]

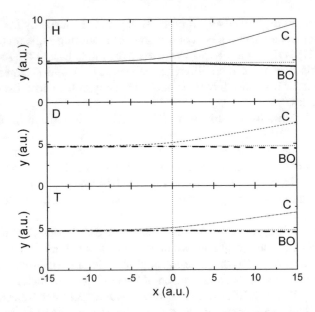

Fig. 5.14 Internuclear trajectories for the system (5.30) calculated with the Coulomb (thin lines) and BO (thick lines) potentials at energy $E = 50$ eV/amu and impact parameter $\rho = 4.68$ a.u. shown in the collision plane with Cartesian coordinates (x, y) in the center-of-mass frame. The trajectories are oriented in such a way that they coincide with each other and with the corresponding straight-line trajectories shown by horizontal thin dotted lines as $x \to -\infty$, which emphasizes the different signs of the scattering angle for the Coulomb and BO trajectories. From [303]

Table 5.1 Scattring angles and geometrical parameters of the Coulomb and BO trajectories calculated at $E = 50$ eV/u and $\rho = 4.68$ a.u., (see Fig. 5.14). From [303]

Target	$V_C(R)$		$V_{BO}(R)$	
	θ (deg)	R_{min} (a.u.)	θ (deg)	R_{min} (a.u.)
H	16.5	5.409	−1.82	4.639
D	9.97	5.106	−1.09	4.655
T	7.76	5.008	−0.85	4.661

sections. Recall here, that radial interactions are considered analytically by searching for hidden crossings (branch points) of the electron energy, which is a multivalued analytic function of the complex internuclear separation R. The amplitudes of radial transitions are averaged over the Stückelberg oscillations and do not depend on the trajectory.

The rotational couplings associated with internuclear axis rotation are treated numerically. They arise in close collisions at small R and are described in the basis of the united-atom hydrogen-like states $\phi_{nlm}(\mathbf{r})$. Rotational interaction induces transitions between states with the same principle n and orbital l quantum numbers, but different projections m of the angular momentum on the internuclear axis.

To describe the evolution of the amplitudes of the states, the time-dependent Schrödinger equation is solved (5.18)–(5.20). The quantity in (5.20) of main interest here is the angle $\varphi(t)$ which defines the internuclear axis orientation in the collision plane. This angle depends on the trajectory, and this is how the trajectory affects the results obtained in the adiabatic approach. In the original ARSENY code [308], the Coulomb internuclear trajectory is used. The Coulomb trajectories in collisions with different targets (5.30) give different time dependencies of the orientation angle $\varphi(t)$, which leads to different rotational coupling contributions to the charge exchange cross sections. The isotope effect corresponding to this mechanism was already described in a previous section. Below we consider another effect of the trajectory caused by using the BO potential instead of the Coulomb one in calculating the angle $\varphi(t)$.

The results of the present calculations of the charge exchange cross sections in collisions of the α-particle with H, D, T targets (5.30) with the use of the BO trajectory are shown in Fig. 5.15 (thick lines) together with our previous results [306] corresponding to the calculations with the Coulomb trajectory in close collisions (thin lines) and results of the END approach [299] (solid symbols). As can be seen in the figure, all the results manifest a strong isotope effect in the energy range under consideration. In the adiabatic approximation, this effect is due to the difference in internuclear trajectories, defined by the same potential, in collisions with targets of different masses. Besides the isotope effect, Fig. 5.15 shows a strong dependence of the adiabatic results on the potential which defines the internuclear trajectory. Figure 5.16 illustrates the difference between the Coulomb and BO trajectories calculated for the following collision parameters: $E = 100$ eV/amu and $\rho = 0.17$. At

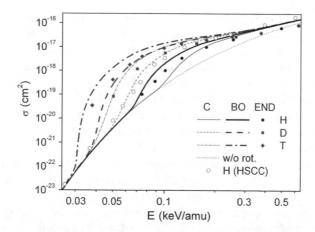

Fig. 5.15 Charge transfer cross sections for the reaction (5.30). Thin and thick lines: present results obtained in the adiabatic approach using the Coulomb and BO internuclear trajectories, respectively. Solid symbols: results of the END approach [312] from [299]. Thin dotted line (the same for all three targets): adiabatic results calculated without rotational couplings. Open circles: results for H from hyperspherical close-coupling calculations [313]. From [303]

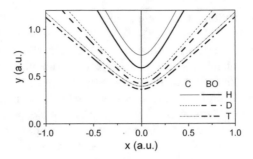

Fig. 5.16 Internuclear trajectories for the system (5.30) calculated with the Coulomb (thin lines) and BO (thick lines) potentials at energy $E = 100\,\text{eV/amu}$ and impact parameter $\rho = 0.17\,\text{a.u.}$ shown in the collision plane with Cartesian coordinates (x, y) in the center-of-mass frame. The trajectories are oriented to emphasize the different values of R_{min}, which has a strong impact on the rotational coupling. From [303]

this parameters, the effect of the trajectory on the cross sections and the rotational couplings that cause this effect are strong. The rotational couplings arises in close collisions (distance R is small), which results in the sensitivity of the cross sections to the distance of closest internuclear approach R_{min}. The smaller R_{min} the larger the effect of rotational coupling and the stronger the dependence on the trajectory. As seen from Fig. 5.16, R_{min} decreases as the target mass increases from H to T, which leads to an increase in the corresponding charge exchange cross section. In the case of the Born-Oppenheimer potential effectively accounting for the electron-nuclei interaction, the values of R_{min} are smaller than that for the Coulomb potential, which leads to a further increase of the cross section.

It can be seen from Fig. 5.15 that the account for the electron-nuclear interaction in adiabatic approximation enlarges the range of the collisional energies where the rotational interaction affects the cross section and improves the agreement of the adiabatic results with the ab initio END results [299]. This is due to the fact that R_{min} is smaller for the BO trajectory than for the Coulomb one for the same energy and impact parameter (see Fig. 5.16), and therefore the BO trajectory describes the rotational couplings at small R more accurately. Improved agreement is achieved at lower energies, while at higher energies the sensitivity of the results to the trajectory is low. The use of the BO trajectory gives better agreement with the END results at energies below $E \approx 0.05\,\text{keV/u}$ for T target, while for H target it happens already at energy $E \approx 0.1\,\text{keV/u}$. This is explained by the fact that a heavier target can enter the region of strong rotational couplings at lower energies.

The adiabatic results corresponding to the calculation without taking into account the rotational couplings, shown in Fig. 5.15 by a thin dotted line, do not depend on the trajectory. Since the effect of the rotational couplings responsible for this dependence decreases at higher energies, all the adiabatic and END results converge to the thin dotted line. At lower energies, where the rotational couplings (5.20) decrease due to the vanishing of the derivative $d\varphi(t)/dt$ as $v \to 0$, the adiabatic results calculated for the different trajectories and targets again coincide with each other. This energy region was not considered in [299], so this prediction of the adiabatic approach cannot be confirmed by comparison with the END results.

The internuclear motion is treated classically in both approaches, so the adiabatic and END results should be the same for the same internuclear trajectory. A good agreement between the BO and END ab initio trajectory used in [299], demonstrated in [303], leads to the conclusion that the difference between the two results is due to an inaccuracy of the adiabatic approximation and/or possible disadvantages of the numerical scheme used to solve the time-dependent Schrödinger equation in [299]. The results of fully quantum hyperspherical close-coupling (HSCC) calculations for the collisions (5.30) with H target [313] are shown in Fig. 5.15 by open circles. The HSCC method based on the solution of the quantum three-body Coulomb problem does not use any approximations, provided that its numerical realization yields converged results. It is noteworthy that the adiabatic results obtained with the BO trajectory are in better agreement with the HSCC results than the END results. Assuming the numerical convergence of the ab initio calculations [299, 313], this means that the adiabatic approach partially compensates for the error caused by the classical treatment of the internuclear motion.

The results of these studies lead to an important consequence: in the implementation of the adiabatic approach, the BO trajectory should be used instead of the Coulomb one, which in fact fully corresponds to the spirit of this approach.

5.8 Resonance Charge Exchange Between Excited States in Slow Proton-Hydrogen Collisions

The study of the resonance charge exchange (RCE) between ground (initial and final) states of the hydrogen atom in slow proton-hydrogen collisions has a long history: starting from the pioneering works where the Firsov [325] and Demkov [326] theory of the RCE was developed, the process was investigated in detail in [327–329]. However, the RCE between excited states can not be described by the existing theory because of their degeneracy. The available data on the RCE process involving excited degenerate states obtained by the close-coupling calculations [313, 330] is very limited, and for energies below 1 a.u. $\approx 27.2\,\text{eV}$ such data is absent. In [323], a theory for the RCE between excited states was developed, extending the Firsov-Demkov theory for the case of degenerate initial and final states.

To illustrate the reliability of the method proposed in [323], we start with the well studied reaction

$$H(n = 1) + H^+ \rightarrow H^+ + H(n = 1), \tag{5.36}$$

using the available data for comparison. Method for calculating the RCE cross section in the low energies collisions (5.36) has been developed by Firsov in 1951 [325]. This method yields the following semiclassical expression for the RCE cross section

$$\sigma = 2\pi \int_0^\infty \sin^2 \left(\int_\rho^\infty \frac{E_g(R) - E_u(R)}{2v\sqrt{R^2 - \rho^2}} R\,dR \right) \rho\,d\rho, \qquad (5.37)$$

where R is the internuclear distance, ρ is the impact parameter, v is the collision velocity and $E_{g,u}(R)$ are the electronic energies corresponded to *gerade* (even) and *ungerade* (odd) states. They are found as the eigenvalues of the two-center Coulomb problem [307] and are shown in Fig. 5.17. The parity of an electronic state characterized by spheroidal quantum numbers n_η, n_ξ and m is defined as $(-1)^{n_\eta}$ [307].

Firsov's theory provides a method for calculating the RCE section, but does not disclose the transition mechanism. In [326], Demkov considered a more general case when the energies of the initial and final states differ by ΔE value, the so-called *quasi*-resonance charge exchange. Based on the exactly solvable Rosen-Zener model [331], he developed a theory widely used in the physics of atomic collisions. In this theory, the electronic states are diabatic in the separated atoms limit, and are adiabatic in the quasi-molecular region. The adiabatic states coincide with the diabatic ones for $R \to \infty$. The electronic transition occurs at the internuclear distance where diabatic (atomic) states rearrange into adiabatic (molecular) ones and the nonadiabatic coupling has its maximum. Thus, a very general mechanism of electronic transitions in the quasi-resonance charge exchange is revealed in the Demkov theory. Importantly, in the limit $\Delta E = 0$ results of this theory are described by (5.37) obtained by Firsov [325].

The one-electron collision system (5.36) allows a much more in-depth analysis due to separability of the two-center Coulomb problem in prolate spheroidal coordinates [307]. As discussed in the previous sections, transitions in slow collisions in the $Z_1 - e - Z_2$ system, where Z_1 and Z_2 are nuclear charges, can be described by the adiabatic theory developed by Solov'ev [28]. It was shown in [332] that in the adiabatic theory transitions due to the Demkov mechanism occur in the so-called P-series of hidden crossings that exists when $|1 - Z_1/Z_2| \ll 1$, that is, in the quasi-resonance case. The P-series was analyzed in detail in [333] and it was found that

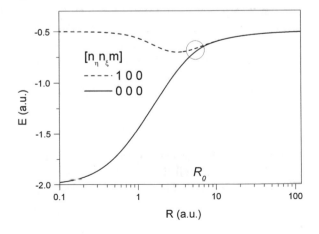

Fig. 5.17 Electronic energies for *gerade* and *ungerade* states participating in the charge exchange process $H(n = 1) + H^+ \to H^+ + H(n = 1)$ as functions of the internuclear distance. The electronic charge exchange transition via the Demkov mechanism occurs in the circled region near the point R_0 defined by (5.38). From [323]

the exchange (underbarrier) coupling between quasi-resonance states in the Rosen-Zener model is described only with exponential accuracy. Nevertheless, taking into account the pre-exponential factor in the exchange interaction [333] leads to almost the same transition probability as in [326].

In the symmetric case $Z_1 = Z_2$, the *gerade* and *ungerade* states are exactly decoupled and the RCE process can not be described by the adiabatic theory [28] because there are no branch points in the complex R plane, connecting the corresponding electronic energy surfaces. As for the Demkov theory, it is applicable to a more general class of collision systems and remains valid even in the resonant case $\Delta E = 0$. A conclusion is drawn (this should be regarded as a probable assumption, and not a proven statement), that the charge exchange transitions in the system (5.36) occur in the regions of rearrangement of atomic states into molecular ones. This happens in the vicinity of a point R_0 (Fig. 5.17) where the corresponding *gerade* and *ungerade* states become approximately degenerate. The value of R_0 can be estimated in the semiclassical approach [307] from the equation

$$R^2 \cdot E(R) + 2\lambda(R) + 4\sqrt{\lambda(R)} = 0, \tag{5.38}$$

where $E(R)$ is the electronic energy and $\lambda(R)$ is the separation constant for the *ungerade* state.

In quantum approach (5.37) is presented as [327, 328]:

$$\sigma = \frac{\pi}{(\upsilon\mu)^2} \sum_{L=0} (2L + 1) \sin^2[\delta_g(L) - \delta_u(L)], \tag{5.39}$$

where $\delta_{g,u}(L)$ are the shifts of the scattering phases, μ is the reduced mass and L is the angular momentum associated with the internuclear axis. Using (5.39), the cross sections for the charge exchange reaction (5.36) were calculated and their energy dependence was studied in many papers beginning with pioneering works [327] and [328]; see, for example, [329]. In this approach, the internuclear motion of colliding particles (protons) were treated classically. To our knowledge, the only fully quantum calculation for a related process was reported in [334] where the quasi-resonance charge exchange between hydrogen in the ground state and deuteron was considered. The calculation was based on the solution of the three-body Coulomb problem.

The Firsov method is used in [323] to calculate the RCE cross sections in collision (5.36). The calculations are performed in three different ways: semiclassical with the cross section defined by (5.37), quantum (5.39) with quantum scattering phase shifts defined from the stationary Schrödinger equation

$$\frac{d^2}{dr^2}\psi_{g,u}(R) +$$
$$\left[2\mu\left(\tfrac{1}{R} + E_{g,u}(R) - \epsilon_{cm}\right) + \tfrac{L(L+1)}{R^2}\right]\psi_{g,u}(R) = 0 \tag{5.40}$$

and quantum (5.39) with semiclassical scattering phase shifts [17]:

$$\delta_{g,u}(L) =$$

$$\int_{R_t}^{\infty} \left[\sqrt{2\mu \left(\epsilon_{cm} - \frac{1}{R} - E_{g,u}(R) \right) - \frac{(L+\frac{1}{2})^2}{R^2}} - k \right] dR$$

$$+ \frac{\pi}{2}\left(L + \frac{1}{2}\right) - kR_t, \tag{5.41}$$

where μ is the reduced mass, ϵ_{cm} is the center-of-mass collision energy, R_t is the turning point and $k = \sqrt{2\mu\epsilon_{cm}}$.

The results for the RCE cross sections [323] converged with respect to all numerical parameters are presented in Fig. 5.18. For comparison, the cross sections obtained in [328] are plotted by dots in the same figure. The cross sections calculated with quantum scattering phase shifts $\delta_{g,u}(L)$ [323] are represented by a solid line. These results are in a good agreement with the results from [328] for all collision energies, with the exception of low energies, where there is an insignificant difference in cross sections. The reason for the difference is the following: in [328], the non-adiabatic corrections to the potential are taken into account while these corrections are not included in the calculations in [323] because of their insignificant influence on the cross sections. The oscillatory structure of the RCE cross sections and locations of the orbiting resonances (angular momentum L of some of them are shown in the figure) are in very good agreement with the corresponding data from [335] and [329]. The region where the calculations with the semiclassical (dashed line) and the quantum (solid line) scattering phase shifts $\delta_{g,u}(L)$ converge is shown in Fig. 5.19 on a larger scale. As seen from the figure, for the energies above 0.017 a.u., where the agreement is very good, the semiclassical phase shifts $\delta_{g,u}(L)$ can be used for the RCE calculation instead of the quantum phase shifts, the calculation of which is very time-consuming. The results obtained using the equation (5.37) (dotted line) do not describe the oscillations of the RCE cross sections and converge with the quantum results obtained with (5.39) at energies above 1 a.u.

Fig. 5.18 Resonance charge exchange cross sections of $H(n = 1) + H^+ \rightarrow H^+ + H(n = 1)$ reaction as a function of the center-of-mass collision energy: solid line—(5.39) with quantum $\delta_{g,u}(L)$, dashed line—(5.39) with semiclassical $\delta_{g,u}(L)$, dotted line—(5.37), dots—experiment [328]. From [323]

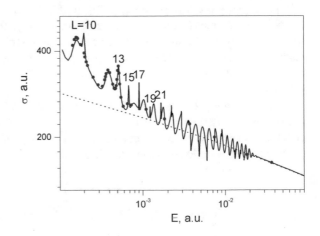

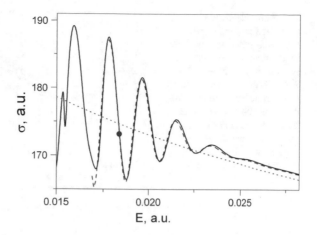

Fig. 5.19 Same as in Fig. 5.18, region of convergence of quantum and semiclassical calculations. From [323]

As an example of RCE between excited degenerate states, the following reaction is considered in [323]

$$H(n = 2) + H^+ \rightarrow H^+ + H(n = 2). \tag{5.42}$$

Six electronic states participate in the reaction (5.42). In the separated-atom limit, $R \rightarrow \infty$, these states converge to the $n = 2$ threshold. Electronic energies for the six states are shown in Fig. 5.20. In the limit of separated atoms, the principle quantum number n is expressed through the spheroidal quantum numbers n_η, n_ξ, and m, identifying an electronic state, as:

$$n = n_\xi + \left[\frac{n_\eta}{2}\right] + m + 1, \tag{5.43}$$

with $[x]$ denoting the integer part of x. In this limit, spheroidal coordinates are transformed into parabolic ones and the correspondence between spheroidal (n_η, n_ξ, m) and parabolic (n_1, n_2, m) quantum numbers is given by [307]

$$n_1 = n_\xi, \ n_2 = \left[\frac{n_\eta}{2}\right]. \tag{5.44}$$

As is clear from Fig. 5.20, the electronic states of the $n = 2$ manifold can be divided into three pairs of *gerade* and *ungerade* states with the same values of n_ξ and m. Within each pair, the value of n_η for the *ungerade* state exceeds that for the *gerade* states by unity, therefore both states have the same values of parabolic quantum numbers (n_1, n_2, m) at $R \rightarrow \infty$. One can notice that degeneracy *within* pairs occur at much smaller R than that *between* pairs. Below, pairs of states defined in this way will be called degenerate. The states constituting the degenerate pair are drawn by lines of the same style in Fig. 5.20.

Fig. 5.20 Electronic energies for *gerade* and *ungerade* states participating in the charge exchange process $H(n = 2) + H^+ \to$ $H^+ + H(n = 2)$ as functions of internuclear distance. The electronic charge exchange transitions via the Demkov mechanism within each degenerate pair occur in the circled regions defined by (5.38). From [323]

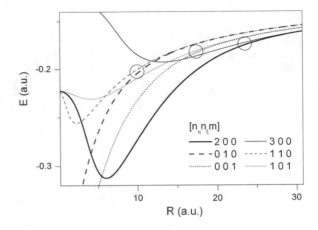

To calculate the cross sections in the RCE process (5.42), the following approach is suggested in [323]: charge exchange transitions within each degenerate pair are considered using the Firsov-Demkov theory, while interactions between the different pairs, as well as with states of other manifolds, are neglected. This approach is justified in adiabatic approximation by the following condition: the circles in Fig. 5.20, indicating the regions of localized charge exchange transitions in degenerate pairs via the Demkov mechanism, are well separated from each other. A similar approach for calculating the quasi-resonance charge exchange cross sections in the process $d\mu(n) + t \to d + t\mu(n)$ between states of the $n = 2$ manifold was used in [336]. The results obtained with allowance for only the Demkov couplings within each quasi-degenerate pair and neglecting all other interactions are in good agreement with the results of fully quantum calculation available for this three-body Coulomb system [337].

Figure 5.21 shows the cross sections of RCE between parabolic states with $n = 2, m = 0, 1$ calculated using approach suggested in [323]. As can be seen from the figure, the RCE cross sections for the transitions between the parabolic states $010-010$ and $001-001$ have a resonant structure and increase with decreasing energy, and the cross section corresponding to the transition $100-100$ does not exhibit resonances and decreases. The potentials $U(R) = \frac{1}{2n^2} + \frac{1}{R} + E_{g,u}(R)$ for three pairs of spheroidal states are plotted in Fig. 5.22. The potentials of 010 and 110 spheroidal states (100 in parabolic coordinates) are purely repulsive. It is possible to analyze the behaviour of the cross section by comparing the positions of the turning point R_t (internuclear distance where the argument of the square root function in (5.41) turns zero) and R_0 (internuclear distance where these states become approximately degenerate). As the energy of collision decreases, the turning point moves to the right and at some point becomes larger than 10 a.u. (R_0 for these states). This happens at an energy of about 0.02 a.u., and when the energy becomes less than this value, the cross section must decrease rapidly. This is indeed the case for the transitions between the parabolic states $100-100$ in Fig. 5.21.

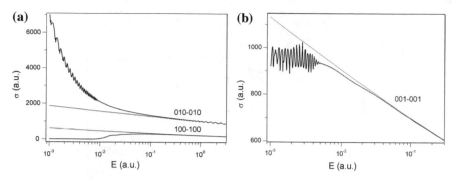

Fig. 5.21 Resonance charge exchange cross sections of $H(n_1, n_2, m) + H^+ \rightarrow H^+ + H(n_1, n_2, m)$ reaction as functions of the center-of-mass collision energy for parabolic states 010, 100 (**a**) and 001 (**b**): solid line—(5.39) with quantum $\delta_{g,u}(L)$, dashed line—(5.39) with semiclassical $\delta_{g,u}(L)$, dotted line—(5.37). From [323]

Fig. 5.22 Potentials $U(R) = \frac{1}{2n^2} + \frac{1}{R} + E_{g,u}(R)$ for *gerade* and *ungerade* spheroidal states as functions of internuclear distance. From [323]

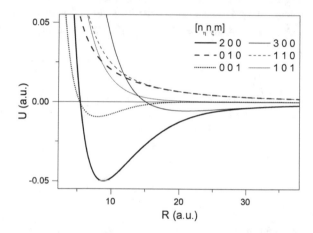

Figure 5.23 shows resonances corresponding to the total angular momentum L in the transition between parabolic states 010$-$010. The potentials

$$U(R) = \frac{1}{2n^2} + \frac{1}{R} + E_{g,u}(R) + \frac{L(L+1)}{2\mu R^2} \qquad (5.45)$$

for spheroidal states 200 and 300 and the angular momentum $L = 90$ are plotted in the insert. The barrier in the 200 state potential causes the first orbiting resonance in Fig. 5.23. The collision energy $E_{L=90}$, where this resonance occurs, is very close to the top of the barrier, which leads to a rather large resonance width.

Figure 5.24a illustrates the convergence of RCE cross sections between 010$-$010 parabolic states obtained with semiclassical and quantum phase shifts $\delta_{g,u}(L)$. The convergence of cross sections defined by (5.37) and (5.39) is shown in Fig. 5.24b. From the figures, one can draw the following conclusion: the semiclassical phase shifts $\delta_{g,u}(L)$ can be used to calculate the RCE cross sections defined by (5.39) when

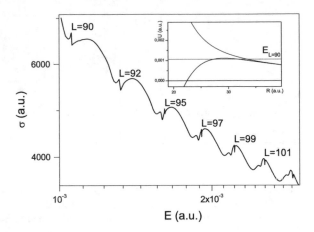

Fig. 5.23 Orbiting resonances in RCE cross sections as functions of the center-of-mass collision energy between parabolic states 010−010. Insert: Potentials (5.45) $U(R) = \frac{1}{2n^2} + \frac{1}{R} + E_{g,u}(R) + \frac{L(L+1)}{2\mu R^2}$, $L = 90$ for 200 and 300 spheroidal states as functions of internuclear distance. From [323]

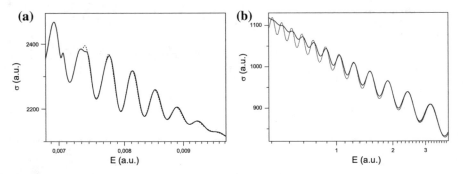

Fig. 5.24 Same as in Fig. 5.21, region of convergence of calculations with quantum and semiclassical $\delta_{g,u}(L)$ (**a**), quantum and (5.37) (**b**). From [323]

the collision energy is above 0.008 a.u., and when the collision energy is higher than 2 a.u., one can use (5.37).

The Clebsch-Gordan coefficients are used to transform parabolic coordinates into spherical [17]. The RCE cross section in transitions between nlm and $nl'm$ spherical states can be written as

$$\sigma_{nlm-nl'm}(v) = \frac{\pi}{(v\mu)^2} \sum_{L=0}(2L + 1) \times$$

$$\left| \sum_{n_1 n_2 m} c(n_1 n_2 m - nlm) A_{n_1 n_2 m}(L) c(n_1 n_2 m - nl'm) \right|^2, \tag{5.46}$$

where

$$A_{n_1 n_2 m}(L) = \sin(\delta_g(L) - \delta_u(L)) \, e^{i(\delta_g(L)+\delta_u(L))} \tag{5.47}$$

is the amplitude of the RCE transition in parabolical coordinates and

$$c(n_1 n_2 m - nlm) = (-1)^{n_1+m} \sqrt{2l+1}$$
$$\times \begin{pmatrix} \frac{n-1}{2} & \frac{n-1}{2} & l \\ \frac{n_2-\bar{n}_1+m}{2} & \frac{n_1-\bar{n}_2+m}{2} & -m \end{pmatrix}. \tag{5.48}$$

The second summation in (5.46) is performed over parabolic states having the same n and m quantum numbers. In [323], the cross sections were calculated for the following RCE transitions: $2s0-2s0$, $2p0-2p0$, $2s0-2p0$ and $2p1-2p1$. The cross sections of the $2s0-2s0$ and $2p0-2p0$ transitions, presented in Fig. 5.25, are equal because of the properties of the Clebsch-Gordan coefficients. The RCE cross section of the $2s0-2s0$ transition for the velocity 0.1 a.u. was calculated in [338] where the semiclassical close-coupling method was used. The result of the calculation is shown by the solid point in Fig. 5.25. The results of the present calculations and calculations performed in [338] differ by about 30%. The difference can be explained in the following way: in the method proposed in [323], other processes such as non resonant charge exchange, excitation and ionization are not taken into account, but the probabilities of these processes become significant when collision energy increases.

The summed cross sections of the $2s0-2s0$ and $2s0-2p0$ transitions [323] together with the total cross section from $2s0$ state at a collision velocity 0.05 a.u., calculated using the six-state molecular close-coupling formalism [330] (solid point) are shown in Fig. 5.26. Compared with the velocity 0.1 a.u., the difference between the two results reduced to 9%. This means that the contribution of the processes unaccounted in [323] decreases when the velocity decreases from 0.1 a.u. to 0.05 a.u. To our knowledge, there are no other data which would lie closer or inside the region of applicability of this theory.

The theory suggested in [323] can not treat the rotational coupling of $2p0$ and $2p1$ states, but a general scaling formula for the $\sigma - \pi$ transition probability in the united-atom limit [339] can be used to evaluate the cross section of the $2p0-2p1$

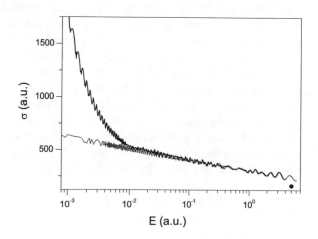

Fig. 5.25 Resonance charge exchange cross sections of $H(n, l, m) + H^+ \rightarrow H^+ + H(n, l, m)$ reaction as functions of the center-of-mass collision energy for spherical states $2s0$ and $2p0$: solid line—(5.39) with quantum $\delta_{g,u}(L)$, dashed line—(5.39) with semiclassical $\delta_{g,u}(L)$, dotted line—(5.37), dot—results from [338]. From [323]

Fig. 5.26 Total resonance charge exchange cross sections of $H(2s0) + H^+ \rightarrow$ $H^+ + H(2s0, 2p0)$ reaction as functions of the center-of-mass collision energy: solid line—(5.39), dotted line—(5.37), dot—data from [330]. From [323]

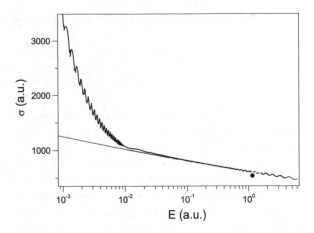

Fig. 5.27 Electronic energies for *gerade* and *ungerade* states participating in the charge exchange process $H(n = 7, m = 0) + H^+ \rightarrow$ $H^+ + H(n = 7, m = 0)$ as functions of internuclear distance

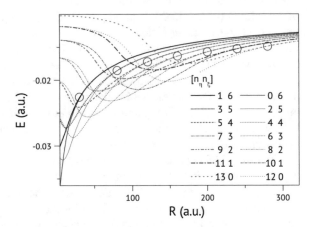

transition caused by this coupling. This formula gives the following estimates: the cross section value changes from 0.1 to 1.85 a.u. over the considered energy region and is small compared with m-conserved transitions. In the present approach, the rotational mixing in the separated-atom limit, which is considered in [340–342], is neglected. This is justified by the fact that, in the adiabatic approximation, the evolution operator mixing the different states of the same n manifold [340] reduces to unity operator in the parabolic basis representation.

The RCE cross section of the $2p1-2p1$ transition in spherical coordinates coincides with the one in parabolic coordinates and is shown in Fig. 5.21b.

To illustrate that the theory, developed in [323] for the case of the $n = 2$ states, is valid for any excited states, the following RCE process is considered:

$$H(n = 7, m = 0) + H^+ \rightarrow H^+ + H(n = 7, m = 0). \tag{5.49}$$

Figure 5.27 shows the electronic energies of the excited degenerate states of hydrogen in the reaction (5.49). As can be clearly seen from the figure, the system of 14 electronic states are split into 7 pairs of the *gerade* and *ungerade* states and the regions of the transitions in each degenerate pair (circles in the figure) are well separated in internuclear distance. This means that the theory [323] can be applied to calculate the RCE cross sections between any excited states at collision energies, where the processes caused by interaction between pairs are negligible.

Chapter 6
Electron Loss Processes

Abstract Electron loss, or *stripping* of projectiles is a competitive charge-changing process to electron capture (Chap. 4), both playing an important role in many fields of physics, for example, in accelerator physics due to their strong influence on ion-beam lifetimes, beam intensities, energy losses (stopping power) and so on. In this chapter, the general features of electron-loss (EL) processes of fast many-electron ions are discussed including the role of the target-density effect, the Bragg's additivity rule and other topics. A special attention is paid to multiple-electron loss (MEL) processes which are of a especial importance at low and intermediate collision energies where they significantly (up to more than 50%) enhance the total EL cross sections.

6.1 Non-relativistic Energies $E < 200$ MeV/u. The Role of Multiple-Electron Processes

Ionization of projectile ion by a target atom or molecule, also called *electron loss*, or *stripping*, is described by reaction:

$$X^{q+} + A \rightarrow X^{(q+m)+} + \sum A + me^-, \quad m \geq 1, \qquad (6.1)$$

where X^{q+} denotes a projectile ion with the charge q, A the target atom (or molecule), m the number of ejected electrons, and $\sum A$ means that the target can be excited or ionized.

The *total* electron-loss (EL) cross section is defined by the sum over all m-values

$$\sigma_{tot}^{EL} = \sum_{m=1}^{N} \sigma_m(\upsilon), \qquad (6.2)$$

where N denotes the total number of projectile electrons, υ the ion velocity, and $\sigma_m(\upsilon)$ the multiple-electron loss (MEL) cross section to ionize m electrons in one collision.

© Springer International Publishing AG 2018
I. Tolstikhina et al., *Basic Atomic Interactions of Accelerated Heavy Ions in Matter*, Springer Series on Atomic, Optical, and Plasma Physics 98,
https://doi.org/10.1007/978-3-319-74992-1_6

One of the most important properties of electron loss is a large (up to 70%) contribution of MEL processes to the total loss cross sections that is significantly pronounced at low and intermediate ion energies. This feature is demonstrated in Table 6.1 where experimental MEL σ_m and total σ_{tot} cross sections are given. As it is seen, even for middle-Z Ar^{8+} ion, having 10 electrons, a contribution of MEL processes with $m \geq 2$ is about 50%. In general, a contribution of MEL processes to the total loss cross sections increases with the target atomic number increasing.

Measurements of the total electron-loss cross sections σ_{tot} have been performed in a wide energy range from a few keV/u up to hundreds of MeV/u in collisions of heavy ions with gaseous targets H_2, He, N_2, O_2, Ne, Kr and Xe (see, e.g., [42, 43, 48, 343–350], and references therein), complex molecules [343], and solid targets (foils) of atoms from Be to U (see, e.g., [351–359]). At low energies, experimental σ_{tot} values increase and reach their maximum approximately at a projectile electron velocity υ_{max} given by

$$\upsilon_{max}^2 \sim 1.5 \, I_P, \tag{6.3}$$

and, then decrease as collision velocity increases. Here I_P denotes the first ionization potential of the projectile in the atomic units of energy (1 a.u. ≈ 27.212 eV). Similar estimate for υ_{max} follows from the plane-wave Born calculations of single-electron loss cross sections [346, 350].

Scaling features of the total electron-loss cross sections are useful for estimations of σ_{tot} in collisions of ions with arbitrary targets in a wide energy range. Figure 6.1 shows a dependence of the scaled experimental total electron-loss cross sections σ_{exp}^{sc} as a function of the Born scaled energy u_B for projectiles from Ar^{6+} to U^{63+} and targets from H_2 to Xe. The following scaling variables are used in the figure:

Table 6.1 Experimental single- and multiple-electron loss cross sections (in 10^{-18} cm^2) of Ar^{8+}, Xe^{18+} and U^{q+}, $q = 28$–42, ions in collisions with noble gases, where σ_1, σ_2, σ_3 and σ_{tot} denote single-, double-, triple- and total electron-loss cross sections

Reaction	Energy, MeV/u	σ_1	σ_2	σ_3	...	σ_{tot}	$\sum_m \sigma_{m \geq 2}/\sigma_{tot}$, %	References
Ar^{8+} + Xe	19.0	23	10	5.5	...	44	48	[104]
Xe^{18+} + He	6.0	3.0	1.7	0.2	...	4.9	39	[343]
Xe^{18+} + Ne	6.0	16	7.8	3.8	...	36	56	[343]
Xe^{18+} + Ar	6.0	24	11	5.6	...	56	57	[343]
Xe^{18+} + Kr	6.0	27	13	7.2	...	75	64	[343]
Xe^{18+} + Xe	6.0	34	16	9.0	...	95	64	[343]
U^{28+} + Ar	3.5	13.4	6.8	4.6	...	40.6	67	[103]
U^{31+} + Ar	3.5	12.5	5.9	3.9	...	34.7	64	[103]
U^{33+} + Ar	3.5	8.7	4.4	3.5	...	26.3	67	[103]
U^{39+} + Ar	3.5	8.0	4.1	2.9	...	19.7	59	[103]
U^{42+} + Ar	3.5	6.7	3.2	2.0	...	13.8	51	[103]

The ratios $\sum_m \sigma_{m \geq 2}/\sigma_{tot}$ show a relative contribution of multiple-electron loss to the total cross sections

Fig. 6.1 Scaled total electron-loss cross sections for heavy ions colliding with atoms as a function of the Born scaled velocity u_B, (6.4). The solid curve denotes a fitted scaled total cross section, see text. From [344]

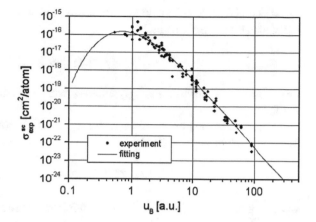

$$\sigma_{exp}^{sc} = \sigma_{tot} \frac{I_P^{1.5}}{u_B^{2.7} Z_T^{1.3}}, \quad u_B = v^2/I_P, \qquad (6.4)$$

where Z_T denotes the target atomic number. Experimental data, shown by solid circles, correspond to energies from 0.7 to 140 MeV/u and taken from measurements performed at Super-HILAC (LLL), LEIR (CERN), Texas A&M cyclotron and GSI, Darmstadt (see [344] for references). The solid curve represents an 'average' scaled cross section, which is described by a sixth-order polynomial formula within 5% uncertainty obtained without relativistic corrections. At high non-relativistic energies, experimental total cross sections are described by the Born asymptotic formula:

$$\sigma_{exp}^{tot} \approx 1.0 \times 10^{-15} \ (\text{cm}^2/\text{atom}) \ \frac{Z_T^{1.3}}{v^2 I_P^{0.5}}. \qquad (6.5)$$

Scaling relations for MEL and total electron-loss cross sections of ions by neutral atoms, are considered in several papers (see, e.g., [360–363]), where cross sections are scaled using a set of atomic parameters such as the projectile ion velocity and nuclear charge, effective number of active electrons and others. There, the scaling relations describe experimental data within a factor of 2–3 and are very useful for practical application.

For calculation of EL cross sections of fast heavy ions, the following approximations are usually used: sudden perturbation method [364], relativistic Born approximation [365], CTMC approach [366, 367] and the energy-deposition (ED) model [368]. At low energies the Born approximation, as a rule, gives strongly overestimated cross-section values, and in this energy range the classical methods for calculation, such as CTMC and ED models, are used.

At energies of $E \approx 50$–500 MeV/u, the total EL cross sections of *light* projectiles are scaled by the Bohr formula:

$$\sigma_{tot}^{EL} \approx Z_T^2 + Z_T, \quad \upsilon \gg I_P^{1/2}, \tag{6.6}$$

i.e., the EL cross sections strongly increase with Z_T, where Z_T is the target atomic number,

In the case of *heavy* projectiles, a scaling law of the total loss cross sections is not so strong as one given in (6.6):

$$\sigma_{tot}^{EL} \approx Z_T^{a(q)}, \quad 1.2 \leq a(q) \leq 1.8, \tag{6.7}$$

where q is the projectile charge. This result follows from calculations by the RICODE and RICODE-M programs, based on the Born approximation (see [43]). Here the exponent $a(q)$ depends on the electronic structure and the charge of the projectile: with the ion charge increasing, the $a(q)$ increases approximately from 1.2 to 1.8. This feature is due to the influence of screening effects in the target atoms: ionization of low-charged ions ($q \sim 1$) occurs at large impact parameters, when the target nucleus is strongly screened by atomic electrons, meanwhile for highly charged ions ($q \gg 1$) the electron ejection from the incident ion takes place at close distances to the target, where the screening effects mentioned are small (see [113]). For example, in [343], where MEL and total EL cross sections were measured for 6-MeV/u Xe^{18+} ions colliding with He, Ne, Ar, Kr, and Xe, a simple liner dependence of σ_{tot} was found, i.e., $a(q) \sim 1$.

A large amount of MEL and total loss cross sections of heavy ions were calculated in the classical approximations such as the CTMC method [366, 367], and the energy-deposition model [368] (see next section). We note that the CTMC calculations have led to significant physical results: a large contribution of multiple-electron loss to the total cross section at low and intermediate collision energies, more gradual dependence of the loss cross sections on the velocity $\sigma_{EL} \sim \upsilon^{-1}$ compared with the Born approximation $\sigma_B \sim \upsilon^{-2}$, a preferential single-electron loss at high non-relativistic energies, and others.

The use of the CTMC method is quite complicated because many electrons and ion trajectories should be taken into account to get enough statistics for the calculated electron-loss or electron-capture cross sections (see Sect. 4.2). The CTMC method is applied to the intermediate collision energy range, where molecular effects can be neglected. Because of the computational difficulties mentioned, the number of publications on CTMC electron-capture cross sections involving heavy ions is quite limited, even for single-electron capture data (see, e.g., [42]).

At high not relativistic energies E, EL cross sections decrease as

$$E \to \infty, \quad \upsilon \to \infty, \quad \sigma_{EL} \to Z_T^2 \ln\upsilon/\upsilon^2 \sim Z_T^2 \ln E/E. \tag{6.8}$$

At *relativistic energies*, EL cross sections do not follow the Born asymptotic law $\ln E/E$ but turns to quasi-constant value due to the influence of relativistic effects (see Sect. 6.4 and [365]). Unlike asymptotic low for ionization of ions by *neutral atoms*, (6.8), electron-loss (projectile ionization) cross sections by *ions* increase asymptotically as $\ln\gamma$:

$$E \to \infty, \quad \upsilon \to c, \quad \sigma_{EL} \sim \ln\gamma \quad \text{for ion–ion impact,} \tag{6.9}$$

$$\sigma_{EL} \approx constant \quad \text{for ion–atom impact,} \tag{6.10}$$

where γ is the relativistic factor and c is the speed of light.

6.2 Energy-Deposition Model for Multiple-Electron Loss of Heavy Ions

6.2.1 Classical Approximation. Basic Assumptions

Classical energy-deposition model (ED) was suggested by Bohr [154] and is based on the assumption that if the kinetic energy $T(b)$, transferred to the projectile electrons by the target atom, exceeds the first ionization potential I_1 of the projectile, i.e., $T(b) \geq I_1$, then ionization occurs with ejection of one or more projectile electrons; here b is the impact parameter.

The ED model was applied in [369] to explain experimental data on multiple-electron ionization of the *atoms* by ions. Later, this model was developed in [368] for ionization of *ions* by atoms, i.e., for calculating electron-loss cross sections. It should be noticed that these two cases of ion-atom collisions are not identical: ionization of an atom by an ion occurs due to interaction of atomic electrons with a long-range Coulomb field of the projectile, whereas ionization of a projectile by an atom is due to interaction of projectile with a field of a neutral target, which is close to the Coulomb interaction at small inter-particle distances and is exponentially small at large distances.

Using the ED model, a DEPOSIT code was created in [368] for calculation of single- and multiple-electron loss cross sections of heavy projectiles by neutral atoms at low and intermediate ion energies. In the code, the kinetic energy $T(b)$, transferred to all projectile electrons, is calculated using the classical Bohr formula [154]:

$$T(b) = \sum_{\gamma} \int \rho_\gamma(r)\, \Delta E_\gamma(p)\, \mathrm{d}\mathbf{r}, \tag{6.11}$$

where ρ_γ denotes the electron density of the projectile γ-th shell at a distance r from its nucleus, ΔE_γ a gain of kinetic energy of projectile electron, interacting with the target, p the impact parameter between the projectile electron and the target nucleus, and the sum over s means summation over all projectile electron shells. The vectors $\mathbf{r}$, $\mathbf{b}$, and $\mathbf{p}$ are related through a simple geometrical expression. The projectile electron density $\rho(r)$ is normalized to the total number N of the electrons:

$$\int_0^\infty \sum_\gamma \rho_\gamma(r)\mathrm{d}r = \int_0^\infty \rho(r)\mathrm{d}r = \sum_\gamma N_\gamma = N. \tag{6.12}$$

In the DEPOSIT code, the electron density $\rho_\gamma(r)$ is calculated using the Slater nodeless functions, and ΔE_γ via the derivative $dU(R)/dR$ of the field $U(R)$ created by the target atom at a distance R from its nucleus. In the code, the analytical expression is used to describe the $U(R)$ with approximation parameters obtained by the Dirac-Hartree-Fock-Slater method for neutral atoms from H to U [370]:

$$U(R) = -\frac{Z_T}{R} \sum_{i=1}^3 A_i \exp(-\alpha_i R), \quad \sum_{i=1}^3 A_i = 1, \tag{6.13}$$

where Z_T denotes the target nuclear charge, A_i and α_i the approximation parameters. The analytical form of the neutral-atom fields (6.13) is very useful for theoretical investigations in many fields of atomic physics and widely used for different applications.

The total electron-loss cross section, i.e., summed over all numbers of ejected electrons, has the form (6.2):

$$\sigma_{tot}(b) \equiv \sum_{m=1}^N \sigma_m = \pi b_{max}^2, \tag{6.14}$$

were the b_{max} is found from equation

$$T(b_{max}) = I_1. \tag{6.15}$$

Therefore, the problem of finding the *total* electron-loss cross sections in the classical ED model is reduced to a calculation of the 3D integral (6.11) (in the CTMC method a few hundred equations should be solved). The calculation accuracy of the ED model is within a factor of 2, i.e., similar to that of the CTMC method. In the ED model there is no limitation on the total number of projectile and target electrons, unlike in the CTMC method, and, moreover, in the ED model the heavier the colliding particles, the more accurate results are obtained.

Experimental total EL cross sections of U^{28+} ions colliding with H_2, N_2, Ar, Kr and Xe targets are shown in Fig. 6.2 in comparison with classical calculations— CTMC and energy-deposition approximations. Experimental data are in a reasonable overall agreement (within a factor of 2) with both treatments except for the H_2 target because for light targets and high collision energies the applicability of the classical approximations is seriously limited.

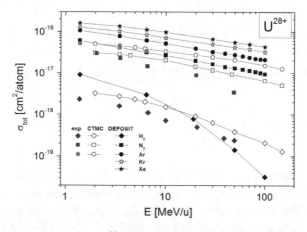

Fig. 6.2 Total EL cross sections of U^{28+} ions colliding with H_2, N_2, Ar, Kr and Xe targets as a function of ion energy. Experiment: H_2 target—solid diamonds, N_2 target—solid squares, Ar target—solid circles, from [95, 96, 101, 102]. Theory: curves with open symbols—CTMC calculations [102], curves with solid symbols (except for Kr target, given by open stars)—DEPOSIT code results. From [368]

Calculations performed by the DEPOSIT code [344] give the following asymptotic behavior of the total EL cross sections on ion energy E:

$$\sigma_{EL} \sim E^{-a(Z_T)}, \quad a(Z_T) \approx 0.8/Z_T^{0.3}, \quad \upsilon^2 \gg I_P, \qquad (6.16)$$

where υ denotes the ion velocity, and coefficients a have the following values for different targets: $a(H) = 0.80$, $a(Ne) = 0.40$, $a(Ar) = 0.34$, $a(Xe) = 0.24$, and $a(U) = 0.21$, i.e., $\sigma_{EL}^{cl} \approx E^{-a}$, $a < 1$. In other words, in the classical approximation the total EL cross sections decrease slower than the Born single-electron cross sections: $\sigma_{EL}^{B} \sim E^{-1}$. These different asymptotic behaviors constitute an important issue in calculation of the total EL cross sections in a wide energy range.

6.2.2 DEPOSIT Code: MEL Cross Sections

As was mentioned before, MEL cross sections for heavy projectile can be very large and therefore, they should be accounted for along with single-electron loss cross sections, especially at law and intermediate collision energies (see Table 6.1). With energy increasing, EL processes take place preferentially with ejection of a single projectile electron (see, e.g., [366]) and are well described by relativistic Born approximation (RICODE-M program) (Sect. 6.4).

In the DEPOSIT code, a probability of ejection m projectile electrons at the impact parameter b is calculated using the statistical Russek-Meli model for m-fold ionization probability $P_m(b, \upsilon$ [371]:

$$P_m(b, \upsilon) = \binom{N}{m} S_m(E_{kin}/I_1) / \sum_{i=1}^{N} \binom{N}{i} S_i(E_{kin}/I_1), \tag{6.17}$$

$$S_m(x) = 2^{\{m-1/2\}} \pi^{\{m/2\}} x^{(3m-2)/2} / (3m - 2)!!, \quad E_{kin} = T(b) - \sum_{i=1}^{m} I_i, \tag{6.18}$$

$$\sum_{m=1}^{N} P_m(b, \upsilon) = 1, \tag{6.19}$$

where $\binom{N}{m}$ denotes the binomial coefficient, $T(b)$ the energy deposited into the projectile by the target atom, (6.11), I_i the i-th ionization potential of the projectile ion, E_{kin} the kinetic energy of m-th ejected electrons, and $\{a\}$ the integer part of a. The binding energies I_i for atoms and ions can be found in [372–375]. The Russek-Meli model has been previously used for studying multiple-electron ionization of neutral *atoms* by protons and positive ions (see, e.g., [369, 376]).

The m-fold and total EL cross sections are calculated by the formulae:

$$\sigma_m(\upsilon) = 2\pi \int_0^{\infty} bdb \, P_m(b, \upsilon), \quad \sigma_{tot}(\upsilon) = \sum_{m=1}^{N} \sigma_m(\upsilon). \tag{6.20}$$

According to (6.19), the Russek-Meli model provides the *unitarity* of the $P_m(b, \upsilon)$ probabilities valid for all impact parameters and all ion velocities υ, therefore, the total EL cross sections σ_{tot}, given by (6.14) and (6.20), are equivalent.

The deposited energy $T(b)$ and multiple-electron loss probabilities $P_m(b)$ calculated by the DEPOSIT code for Xe^{18+} ions colliding with Xe atoms at energy of E = 6 MeV/u are shown in Fig. 6.3(left and right) as a function of impact parameter b. The left figure shows that the curve $T(b)$ intersects with the first ionization potential I_P of Xe^{18+} at $b_{max} \sim 1.2 \, a_0$ (a_0 is the Bohr radius), which is the maximum impact parameter contributing to the loss process: collisions with $b > b_{max}$ do not contribute because they cannot ionize even a single electron from Xe^{18+} ion.

As also seen from the Fig. 6.3, left, that at small impact parameters $b < 0.5 \, a_0$, the deposited energy is transferred mainly to the inner-shell electrons of Xe^{18+} resulting in multiple-electron loss, $m \gg 1$, as seen on the right figure. This is a clear indication of importance of ejection of the projectile inner-shell electrons in EL process.

Figure 6.4 shows experimental [343] and calculated MEL cross sections for Xe^{18+} ions bz inert gases He, Ne, Ar and Xe at 6 MeV/u. It is seen an overall agreement between theory and experiment, except for light atoms (He) because for projectile ionization of light ions, the classical approximation tends to fail, but it woks quite well for heavy atoms where the Born approximation fails (see [377–382]).

A comparison of MEL cross sections, calculated by the classical method—CTMC method [240] and the energy-deposition model [377]—for U^{28+} + Ar collisions as a function of ion energy is shown in Fig. 6.5. At relatively low energies $E = 3$–10 MeV/u, an agreement between both models for σ_m for $m < 10$ is quite good, whereas

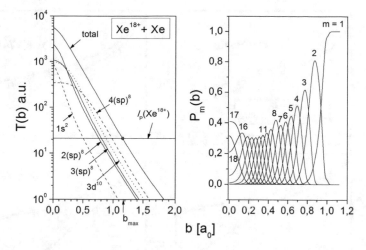

Fig. 6.3 Electron loss of Xe^{18+} ions by Xe atoms at collision energy of $E = 6$ MeV/u ($v = 15.5$ a.u.). Left: deposited energies $T(b)$ (in a.u.) to different electron shells of Xe^{18+} and the total energy as a function of impact parameter b, DEPOSIT code. The horizontal line $I_P = 21$ a.u. (572.5 eV) corresponds to the first ionization potential of Xe^{18+} showing the minimum energy deposit required for ionization. Right: multiple-electron probabilities $P_m(b)$ for ionization of Xe^{18+} colliding with Xe atom. Calculations by the DEPOSIT code from [344]

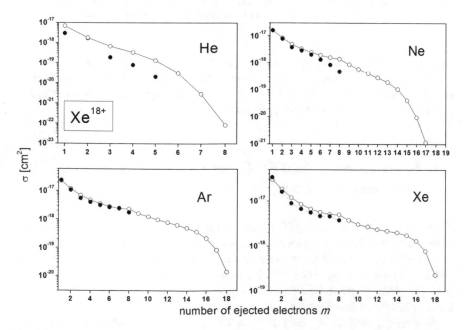

Fig. 6.4 Multiple-electron loss cross sections of Xe^{18+} ions colliding with He, Ne, Ar and Xe at 6 MeV/u: experiment—solid circles [343], theory—open circles, DEPOSIT code. From [377]

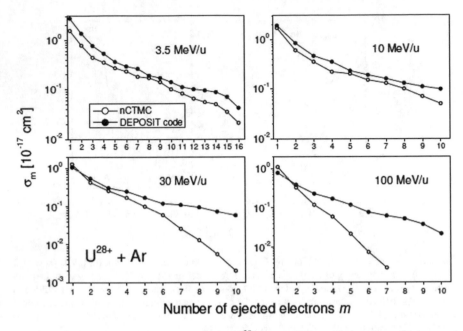

Fig. 6.5 Multiple-electron loss cross sections of U^{28+} ions colliding with Ar at low to high energy range. Open circles—CTMC results [240], solid circles—DEPOSIT code. For large m values, discrepancy between two models increases with collision energy increasing. From [377]

at energies $E = 30$–100 MeV/u the discrepancy is poorer, except for ejection of $m < 4$ electrons. We note, that since electron-loss processes are dominated by losses of $m = 1$, 2 and 3 electrons, this discrepancy is not very important when the total loss cross sections are of interest for applications.

6.3   Recommended Total EL Cross Sections at Non-relativistic Energies

Experimental data, classical (CTMC method and ED model) and the Born calculations of the total EL cross sections show that at high non-relativistic energies, the main contribution to the total loss cross section σ_{tot} is given by single-electron processes and σ_{tot} decreases as $\sigma_{tot} \approx ln E / E$ according to the Born approximation (see, e.g., [11, 42, 346]). At relativistic energies, σ_{tot} shows a quasi-constant behavior due to relativistic effects (see Sect. 6.4).

At present, there is no one theory that can reproduce a behavior of EL cross sections over the whole collision energy range: at low and intermediate energies the classical approach is used, and the Born approximation is applied at high energies. In order to present EL cross sections in a wide energy range, a semiempirical procedure is suggested in [368] to match cross sections at low, intermediate and high energies and to present 'recommended' data using a simple formula:

$$1/\sigma_{rec} = 1/\sigma_{DEPOSIT} + 1/\sigma_{RICODE}, \qquad (6.21)$$

where σ_{rec} denotes recommended EL cross section, $\sigma_{DEPOSIT}$ the total cross section calculated by the DEPOSIT code, and σ_{RICODE} the single-electron loss cross section, calculated by the RICODE or RICODE-M program (latest version of RICODE) at relativistic collision energies. A formula similar to (6.21) is previously used in [378] for *electron-impact* ionization collisions to match cross sections at very low and very high electron energies.

Experimental total EL cross sections of U^{10+} and U^{28+} ions, colliding with gaseous targets as a function of ion energy, are shown in Fig. 6.6 in comparison with calculations performed by the DEPOSIT and RICODE-M programs—'recommended' data (see [379] for detail). Electron-loss cross sections increase with the target atomic number, decrease with ion energy increasing and then saturate due to relativistic effects (see below).

Figure 6.7 shows calculated total electron-loss cross sections for W^{q+} ions colliding with H and He atoms. These data are required for applications in high-power

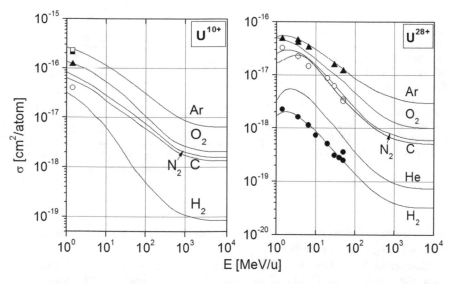

Fig. 6.6 Total electron-loss cross sections of U^{10+} and U^{28+} ions colliding with gaseous targets as a function of energy. Left: U^{10+} projectile. Experiment: H_2 target—open circle [96]; N_2—solid circle [96] and solid triangle [361]; Ar—open square [361]. Right: U^{28+} projectile. Experiment: H_2 target—solid circles [96, 99, 100, 103]; N_2—open circles [96, 99, 100, 103]; Ar—solid triangles, [95, 103]. Theory: solid curves on both figures—recommended data (DEPOSIT + RICODE-M result), [379]. From [379]

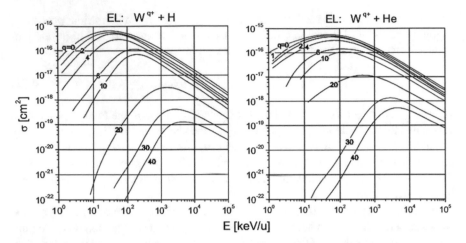

Fig. 6.7 Calculated 'recommended' total electron-loss cross sections of W^{q+}, ions $q = 0$–40, in collisions with H and He as a function of ion energy—DEPOSIT + RICODE result, (6.21). From [302]

Fig. 6.8 EL and EC cross sections of 11 MeV/u U^{q+} ions colliding with H_2, He, N_2 and Ar atoms as a function of uranium ion charge q. Experiment: solid circles—EL and solid triangles—EC for He, $q = 60$–70 [107]. Theory: solid curves—combined DEPOSIT and RICODE electron-loss calculations and dashed curves—EC cross sections, CAPTURE code, from [380]. From [380]

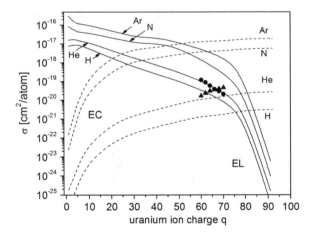

fusion plasma devices because tungsten is used as a most preferable material for plasma-facing components (see [302]).

Calculated electron-capture (EC) and electron-loss (EL) cross sections of uranium ions in collisions with H, He, N and Ar atoms at ion energy of 11 MeV/u as a function of uranium charge states are shown in Fig. 6.8 in comparison with recent RIKEN experimental data for He [107]. As seen from the figure, EC cross sections increase monotonically with increasing the ion charge q as all theories predict. A q-dependence of EL cross sections is different for different q: at small and intermediate q values, $q < 40$, corresponding to heavy many-electron systems, EL cross sections decrease quite slowly due to contribution of ionization from inner-shell electrons of uranium ions. At high q, i.e., for highly charged ions, process clearly shows a different dependencies on the ion charge, and EL cross sections decrease very rapidly for $q >$ 70. The figure clearly shows that for 11 MeV/u uranium ions, the best strippers are the lightest atoms—H and He.

Experimental multiple-electron and total loss cross sections of U^{28+} ions, colliding with the main accelerator residual-gas components H_2, N_2 and Ar, are shown in Fig. 6.9 as a function of ion energy in comparison with combined calculations performed by the DEPOSIT and RICODE-M programs (see [379] for detail). A contribution of MEL processes decreases with the ion energy, and at relativistic energies $E > 300$ MeV/u the main contribution to the total cross sections is given by single-electron loss processes.

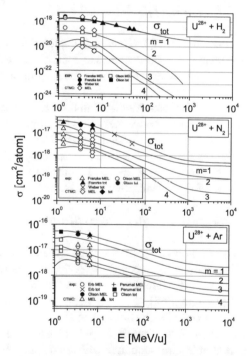

Fig. 6.9 Multiple-electron and total loss cross sections of U^{28+} ions colliding with H_2, N_2 and Ar as a function of ion energy. **H_2 target**: Experiment: open and solid circles—MEL and total cross sections [96]; open and solid squares—MEL and total cross sections at 3.5 and 6.5 MeV/u, [103]. CTMC calculations: diamonds—MEL data for $m = 1, 2, 3$ [103]. **N_2 target**: Experiment: open and solid triangles—MEL and total cross sections, [96]; open and solid circles—MEL and total cross sections at 3.5 and 6.5 MeV/u, [103], crosses—total cross sections [99, 100]. CTMC calculations: open and solid diamonds—MEL and total data [362]. **Ar target**: Experiment: open and solid squares—MEL and total cross sections [95], open and solid circles—MEL and total cross sections at 3.5 and 6.5 MeV/u, [103]. CTMC data: crosses, [362]. In both figures, solid curves—DEPOSIT + RICODE-M result, (6.21), [379]. From [379]

6.4   Electron Loss at Relativistic Energies

At relativistic energies $E \geq 200\,\text{MeV/u}$, experimental and theoretical data on single-electron loss cross sections are quite limited and available mainly for few-electron ions [289, 383–387].

A general treatment for single EL cross sections of an *arbitrary* projectile and a target atom is described in [387] using a relativistic Born approximation with account for magnetic interactions between colliding particles. Based on this treatment, two computer programs RICODE and RICODE-M (Relativistic Ionization CODE) were developed (see [113, 365], respectively) in the momentum-transfer representation. Unlike the RICODE, the RICODE-M uses relativistic wave functions for the active projectile electron in the bound and continuous states (see below).

In the relativistic Born approximation, the ionization matrix element M_{if} has the form [289]:

$$M_{if} = \langle f|(1 - \beta\alpha_z)e^{i\mathbf{Q}\cdot\mathbf{r}}|i\rangle, \tag{6.22}$$

where Q denotes the momentum transfer, $\beta = v/c$ the relativistic factor, c the speed of light, α_z the z-component of the Dirac matrix α, r the distance from the projectile nucleus, $|i>$ and $|f>$ the total wave functions of the system in the initial and final states.

The first term in (6.22) is the 'usual' non-relativistic matrix element, describing the electric (Coulomb) interactions between colliding particles, whereas the second term describes the *magnetic* interactions between them. Calculation of the ionization matrix elements M_{if} with both terms constitutes quite complicated problem which was solved before only for ionization of H- and He-like ions (see, e.g., [289]).

An order of magnitude of the second (relativistic) term in (6.22) can be estimated as

$$\beta\alpha_z \sim v/c \cdot <p_e>/m_e \sim v/c \cdot v_e/c, \tag{6.23}$$

where m_e, v_e and $< p_e >$ denote mass, orbital velocity and impulse matrix element for the active projectile electron. As is seen from (6.23), the relativistic interaction is the largest ($\beta\alpha_z \sim 1$) when both velocities v and v_e are close to the speed of light c.

In the RICODE-M program [365], the relativistic interaction (6.22) is used and *relativistic* radial wave functions for the bound and ejected projectile electrons. The relativistic radial wave functions are found by solving the Schrödinger equation with the effective field of the atomic core calculated with relativistic Dirac wave functions.

The main goal of the RICODE-M program is to calculate *single-electron* loss cross section σ_{EL} for *arbitrary* heavy many-electron projectiles colliding with arbitrary atoms and ions. In the RICODE-M, the EL cross section $\sigma_{EL}(v)$ has the structure [365, 387]:

$$\sigma_{\text{EL}}(v) = \frac{8\pi a_0^2 N_{nl}}{v^2} \int_{q_0}^{\infty} Z_T^2(Q)\frac{dQ}{Q^3} \left(|F(Q)|^2 + \frac{\beta^2(1 - Q_0^2/Q^2)}{(1 - \beta^2 Q_0^2/Q^2)^2}|G(Q)|^2 \right), \tag{6.24}$$

$$v = \beta c, \quad Q_0 = (I_{nl} + \epsilon)/v, \quad (6.25)$$

where $a_0 \approx 0.5292 \times 10^{-8}$ cm denotes the Bohr radius, v the ion velocity, n and l the principal and orbital quantum numbers of the projectile electron shell with ionization potential I_{nl} and N_{nl} number of equivalent electrons, respectively, ϵ the energy of ejected electron, $Z_T(Q)$ the effective charge of the target, which does not coincide with the target nuclear charge and, in general, depends on Q and the minimal momentum transfer Q_0. The term proportional to $|F|^2$ is used for non-relativistic collisions, meanwhile the one proportional to $|G|^2$ takes into account the relativistic (magnetic) interactions.

As an example of the influence of relativistic effects on the radial wave function, an electron density of the 7s-orbital in a super-heavy neutral Rg atom (nuclear charge $Z = 111$, configuration $6d^9 7s^2$), calculated by different approaches, is shown in Fig. 6.10. The dotted curve represents a fully non-relativistic calculation, i.e., with non-relativistic binding energy and non-relativistic core potential $U_c(r)$. The dashed curve is a result of the RICODE program with relativistic binding energy but with non-relativistic potential $U_c(r)$, and the curve with open circles corresponds to the RICODE-M calculations using both relativistic core potential and relativistic binding energy. All three curves are compared with relativistic coupled-cluster calculations based on the Dirac-Coulomb-Breit Hamiltonian [388]. As it is seen from the figure, the RICODE-M result is very close to the fully relativistic calculations [388], especially concerning the main maximum of the radial wave function. The relativistic effects increase the maximum value of 7s-electron density by a factor of 1.5 and

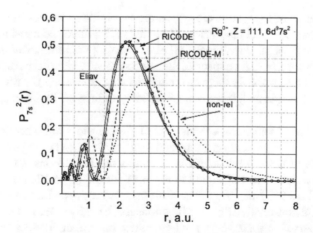

Fig. 6.10 Calculated electron density $P_{7s}^2(r)$ of the 7s-orbital in neutral super-heavy Rg atom with the nuclear charge $Z = 111$. Solid curve—fully relativistic calculation [388]; dashed curve—the non-relativistic wave function but relativistic binding energy, RICODE result; solid curve with open circles—relativistic wave function and relativistic binding energy, the RICODE-M result; dotted curve—fully non-relativistic calculation (see text). From [365]

Table 6.2 Calculated relativistic (rel), calculated by the RICODE-M code, and non-relativistic (non-rel) binding energies ε (a.u.) of neutral Rg atom ($Z = 111$), having electronic configuration $1s^2 2s^2 ... 5d^{10} 6s^2 6p^6 5f^{14} 6d^9 7s^2$

Shell	ε_{rel} HF [373]	ε_{rel} [365]	$\varepsilon_{non\text{-}rel}$ [365]
$7s_{1/2}$	0.4276	0.4278	0.2441
$6d_{5/2}$	0.4119	0.4118	0.6477
$6d_{3/2}$	0.5172	0.5171	0.6477
$6p_{3/2}$	2.2765	2.2764	2.3227
$6p_{1/2}$	3.8476	3.8477	2.3227
$6s_{1/2}$	5.3549	5.3564	3.8094
$5f_{7/2}$	3.0226	3.0224	2.7984
$5f_{5/2}$	2.7986	2.7984	2.7984
$5d_{5/2}$	10.1982	10.1979	11.0519
$5d_{3/2}$	11.2795	11.2791	11.0519
$5p_{3/2}$	17.2870	17.2865	16.6413
$5p_{1/2}$	24.6863	24.6874	16.6413
$5s_{1/2}$	28.7578	28.7645	19.7148
$\cdots$	$\cdots$	$\cdots$	$\cdots$
$1s_{1/2}$	6898.68	6900.22	5481.18

The data are compared with relativistic Hartree-Fock calculations [373]

shift it towards the nucleus. As is seen below, the influence of relativistic effects on the wave functions leads to a change in electron-loss cross sections in maximum of about 30–40%.

The influence of the relativistic effects on the binding energies in neutral super-heavy Rg atom ($Z = 111$) is illustrated in Table 6.2, where the non-relativistic and relativistic results obtained by the RICODE-M are compared with relativistic Hartree-Fock calculations [373]. As expected, the effects are very significant (a factor of 1.5) for ns orbitals because only the wave functions $R_{ns}(r) = P_{ns}(r)/r$ for ns states are non-zero at the origin and, therefore, are strongly influenced by interaction with the nucleus.

Some results of numerical calculations of electron-loss cross sections of heavy and super-heavy many-electron atoms and ions are presented in Figs. 6.11 and 6.12. The influence of the relativistic effects in the case of collisions between Rg ($Z = 111$) and He atoms is demonstrated in Fig. 6.11 for the projectile ionization of and outer $7s$- and inner $6d$-electrons. The difference in cross-sections due to the use of relativistic and non-relativistic wave functions for the active electron is shown (solid and dashed curves, respectively). For both cases, this effect is rather small ($\lesssim 10\%$) although the effect is a little larger for electron-loss cross section with ejection of $6d$-electron. But the influence of the relativistic interactions between the colliding Rg and He atoms is very significant (~ 40–50%) for both $7s$- and $6d$-electrons shown in the figure by difference in the cross sections labeled V_{rel} and $V_{non\text{-}rel}$. In the case

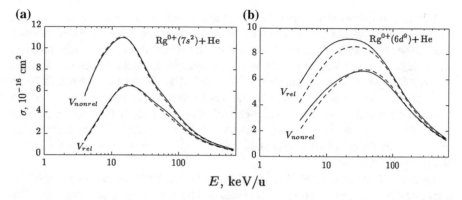

Fig. 6.11 Electron-loss cross sections of neutral Rg atom (nuclerar charge $Z = 111$) colliding with He atom with ejection of $7s$ (left) and $6d$ (right) electrons as a function of ion energy. Dashed curves—calculations with the non-relativistic wave functions for the bound and continuum states, the RICODE program [113]; solid curves—with relativistic wave functions, the RICODE-M program [365]. V_{rel} and $V_{non-rel}$ show cross sections calculated using relativistic and non-relativistic interactions, respectively. From [365]

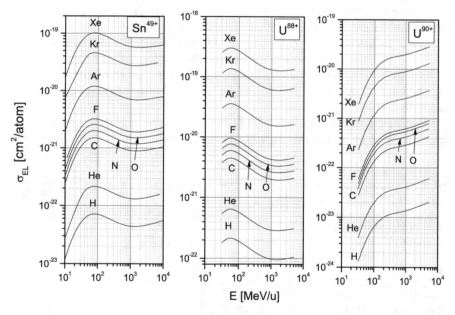

Fig. 6.12 Calculated total electron-loss cross sections of uranium and tin ions by residual gas atoms H, He, C, N, O, F, Ar, Kr and Xe: results of RICODE-M program. From [389]

of ejection of $7s$-electron, the relativistic interaction leads to a decrease of the cross section, meanwhile the electron-loss cross section for $6d$-electron is increased.

Calculations show [365]) that the use of relativistic wave functions changes the electron-loss cross-sections of neural atoms and low-charged ions by about 20–30% around the cross-section maximum compared to those calculated with non-relativistic wave functions; for electron loss of highly charged ions, the influence of using the relativistic wave functions is rather small—about 10%. At present, the RICODE-M can provide the most accurate data on single-electron loss cross sections for heavy many-electron ions at high ion energies including the relativistic domain. At low and intermediate energies, it is necessary to take into account multiple-electron loss cross sections which strongly contribute to the total EL cross sections even for highly charged projectile ions.

Figure 6.12 shows the result of calculations by the RICODE-M program of the total electron-loss cross sections of uranium and tin ions by gaseous targets H_2, He, C, N_2, O_2, F, Ar, Kr and Xe at relativistic energies $E > 400$ MeV/u [389]. These data are required to estimate lifetimes of relativistic heavy-ion beams in a new High Energy Storage Ring (HESR), GSI, Darmstadt. As seen from the figure, the EL cross sections shows the approximate dependence $\sigma_{EL} \sim Z_T^2$ where Z_T is the target atomic number.

In Fig. 6.13 all available experimental data and theoretical calculations of the total EL cross sections for U^{28+} ions colliding with Ar are shown as a function of ion energy to show the influence of various approximations on the EL cross sections. The cross section, marked *non-rel.*, is the result of the LOSS code [350] in a pure Born approximation, i.e., using both a non-relativistic interaction and wave functions. This method leads to the Born asymptotic behavior: $\sigma_B \approx \ln E/E$ and an overestimation of experimental data by up to a factor of 2 at $E \sim 2$ MeV/u, where the cross section shows its maximum.

Classical CTMC cross sections give quite good description of experimental data at intermediate energies, including maximum, but overestimates them at high energies

Fig. 6.13 Total EL cross sections for U^{28+} + Ar collisions as a function of ion energy. Experiment: solid triangle up [95], solid triangle down [103], open square [101], solid circles [100]. Theory: dashed line—CTMC result [102], non-rel. Non-relativistic result by the LOSS code [350], recommended data—DEPOSIT + RICODE-M result. From [365]

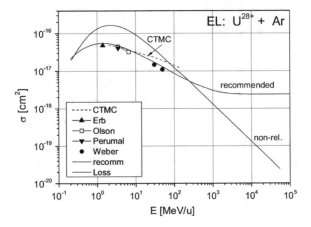

because the CTMC method does not take into account the relativistic interaction between colliding particles. The best agreement between theory and experiment is achieved by *recommended* cross sections [(DEPOSIT + RICODE codes, (6.21)], which agree well with experiment in the intermediate energies and turn to the constant value at relativistic energies. Unfortunately, experimental electron-loss cross sections are available only at non-relativistic energies $E < 200$ MeV/u, therefore, it would be very important to perform such measurements in the energy range, e.g., between 200 MeV/u and 10 GeV/u which will be possible in the framework of the new FAIR project [51]. In general, the quasi-constant behavior of the electron-loss cross sections is the most striking feature of the loss processes of heavy ions by neutral atoms at relativistic energies.

6.5 Bragg's Additivity Rule for EL Cross Sections

Information about multiple and total electron-loss cross sections for heavy ions colliding with molecular targets is highly required for many applications but is limited in both experimental and calculated data (see, e.g. [43, 109, 110, 343, 361, 363]). For example, these data are needed to estimate the vacuum conditions in accelerator machines where molecules like H_2, H_2O, CO_2, CH_4 etc. constitute an important part of the residual-gas components in vacuum systems (see [224, 379]).

Calculations of electron-capture (and electron-loss) cross sections even for a "simple" molecular target H_2 meet big difficulties as it is seen from calculations using the absorbing sphere model [229], the electron tunneling model [390] or a classical phase-space model [391, 392]. Therefore, to interpret collisions of fast ions with molecules, the *Bragg's additivity rule* is usually used, in accordance with which the interaction cross section for a molecule is presented as a sum of the cross sections for atoms in the molecule:

$$\sigma_{mol} = \sum_i n_i \sigma_i(Z_i), \tag{6.26}$$

where n_i is the number of atoms with atomic number Z_i. For example, the electron-loss (or capture) cross section of an ion colliding with CO_2 molecule is presented as: $\sigma_{EL}(CO_2) = \sigma_{EL}(C) + 2\sigma_{EL}(O)$. The Bragg's additivity rule quite often (not always) provides a reasonable agreement between theory and experiment depending on the projectile charge and velocity (see Sect. 4.4, devoted to the Bragg's rule for electron capture).

The additivity rule was examined experimentally in the work [343] by comparing the measured partial σ_m and total σ_{tot} EL cross sections of 6 MeV/u-Xe^{18+} ions colliding with noble-gas atoms and various molecular targets: H_2, CH_4, C_3H_8, SiH_4, N_2, CO, CO_2, O_2, C_3H_8, CF_4, and CF_6. It was found that under experimental conditions, the additivity rule works very well for the measured total and many-fold EL cross sections showing that the target molecule acts as an ensemble of individual atoms in the loss processes, which is illustrated in Fig. 6.14. There, the total EL cross sections σ_{mol} are shown as a function of the target average atomic number $\overline{Z}$:

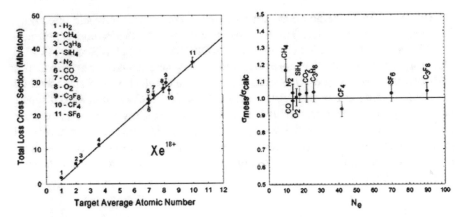

Fig. 6.14 Electron loss of 6-MeV/u-Xe^{18+} projectiles stripped in molecular targets. Left: total electron-loss cross sections per atom as a function of target average atomic number, (6.27). The solid line is fitted through experimental data points from He to Ne. Right: ratio of the measured total EL cross sections and calculated ones using the additivity rule, (6.26), as a function of the total number of the target electrons N_e. From [343]

$$\overline{Z} = \sum_i Z_i n_i / N, \, \sigma_{mol} = N\sigma(\overline{Z}), \qquad (6.27)$$

where N denotes the total number of atoms in the molecule, and $\sigma(\overline{Z})$ the cross section for an *atom* with atomic number $Z_i = \overline{Z}$. The solid line corresponds to experimental data obtained for atomic targets from He to Ne. It is seen that the EL cross sections increase linearly with $\overline{Z}$ and are very close to the data measured for atomic targets.

The validity of the additivity rule for the total EL processes is clearly seen in Fig. 6.14, right, where ratios of the measured and calculated EL cross sections using (6.26) are shown as a function of the total number of molecular electrons. All data, except H_2 and CH_4 molecules, are described by the additivity rule within 6% uncertainty or better.

In [343] it was also demonstrated the validity of the additivity rule for the many-fold EL cross sections as shown in Fig. 6.14 (left and right figures), where experimental cross sections for molecular targets are compared with the data for atomic targets having close atomic numbers. There is a small dependence of the cross sections on the number of atoms per molecules except for cases with a large number of ejected electrons $\Delta q \sim 6$–8, which is more probably related to collisions with small impact parameters.

In general, the additivity rule can be applied to EL cross sections of heavy ions at rather high energies when molecular effects, such as molecular bonding and others, can be neglected and the molecule can be treated as an ensemble of individual atoms.

6.6 Influence of the Target-Density Effect on Electron-Loss Processes

The influence of the target-density (gas-solid) effect on electron-capture cross sections, discussed in Sect. 4.3, is found to be very large (up to more than one order of magnitude) leading to a strong *decrease* of the capture cross sections depending on the relative velocity and atomic structure of colliding particles. In contrast, the density effect (DE) in electron-loss processes results in an *increase* of the loss cross sections by a factor of 2–3.

Let us consider single-electron loss process, i.e., projectile ionization by the target atom in the form:

$$X^{q+}(n_0) + A \rightarrow X^{(q+1)+} + A + e^-, \tag{6.28}$$

where n_0 is the principal quantum number of the projectile ground state, and A is the target atom. As the target density increases, more projectile ions begin to be excited into n-states due to excitation by the target atoms, so the total loss cross section $\sigma_{ion}^{DE}(v)$ with the density effects included can be written in the form [182]:

$$\sigma_{EL}^{DE}(v) = \sigma_{EL}(n_0) + \sum_{n \geq n_0} \sigma_{ex}(n_0 - n)\, B(n), \tag{6.29}$$

$$B(n) = \frac{1}{1 + A(n - n_0)/\rho_T v \sigma_{EL}(n)}, \tag{6.30}$$

where ρ_T denotes the target density, $B(n)$ the branching-ratio coefficient of the excited state n, $\sigma_{EL}(n_0)$ electron-loss cross sections from the ground state n_0 without account for the DE, $\sigma_{ex}(n_0 - n)$ excitation cross section into n state, arising in collisions with the target atom, and $A(n - n_0)$ the total radiative transition probability from the excited state n into the ground state n_0. As a rule, the largest contribution to the sum in (6.29) is given by transition from the ground state into resonance (nearest optically-allowed) level n_r with the largest excitation cross section. Equations (6.29) and (6.30) show the main dependencies of the loss cross section on the target density and other atomic parameters, which can easily be generalized to the case of ionization from levels with the orbital quantum numbers nl [182].

Equation (6.30) show that the influence of the density effect is large if the ionization rate from excited state is larger than the radiative decay rate from the same level:

$$\rho_T v \sigma_{EL}(n_r) \gg A(n_r - n_0), \tag{6.31}$$

where n_r is the principal quantum number of the resonance level.

If a target density is small (a dilute gas), $\rho \rightarrow 0$, the branching ratio coefficient $B \rightarrow 0$, and the electron-loss cross section is given by the 'usual' formula for binary collisions:

$$\sigma_{EL}^{DE}(v) \approx \sigma_{EL}(n_0), \quad \rho_T \rightarrow 0, \quad B \rightarrow 0. \tag{6.32}$$

In the opposite case of a very dense target (solid state), the coefficient $B \to 1$, and the loss cross section is approximately defined by the sum of electron-loss cross section from the ground state and excitation cross section from the ground state into the resonance one:

$$\sigma_{EL}^{DE}(v) \approx \sigma_{EL}(n_0) + \sigma_{ex}(n_0 - n_r), \quad \rho_T \to \infty, \quad B \to 1. \tag{6.33}$$

Since

$$\sigma_{EL}(n_0) \approx \sigma_{ex}(n_0 - n_r), \tag{6.34}$$

equation (6.33) reads that the electron-loss cross section of ions in a dense target is roughly two times larger than that in the low-density target.

Let us recall that, under the conditions of electron capture, the influence of the DE is more significant on the cross section and can lead to its reduction by more than 10 times. The mutual changes in electron-loss and capture-cross sections with the target density give a qualitative explanation to an increase of the mean charge for ion beams passing through dense media (see Sect. 4.3).

Figure 6.15a demonstrates the influence of the target-density effect on electron-loss and capture cross sections for collisions of 6 MeV/u-Ar^{8+} ions with carbon foil of the density $\rho_T \approx 1.1 \times 10^{23}$ cm^{-3} as a function of exit-ion charge q. First of all,

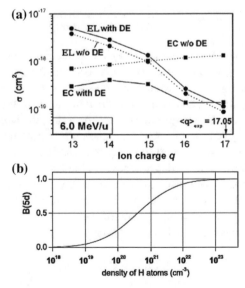

Fig. 6.15 a Influence of DE on single-electron loss (EL) and capture (EC) cross sections in collisions of 6 MeV/u-Ar^{8+} ions with carbon foil as a function of the exit-ion charge q calculated with (solid curves) and without (marked w/o, dotted curves) the DE. Crossings of EL and EC curves show approximate mean charge value $<q>$. Experimental $<q>$-value is $<q>_{exp} = 17.05$. From [393]. **b** Calculated branching-ratio coefficient $B(5d)$ for 1.4 MeV/u-U^{28+} collisions with atomic hydrogen as a function of hydrogen density. From [255]

it is seen that the DE is very significant for EC cross sections (about one order of magnitude at $q = 17$), whereas the effect is about a factor of 1.5 for EL cross section. Secondly, the inclusion of the DE leads to a much better agreement with the average mean charge: $<q> \approx 15$ if calculated without DE and $<q> \approx 16.8$ obtained with the DE accounted for, the experimental value $<q>_{exp}$ is 17.05.

In some cases, the influence of the DE on EL cross sections can be very small, especially, in collisions of heavy ions with light target atoms. In Fig. 6.15b, a branching-ratio coefficient, calculated in the relativistic approximation for excited $5d$-state of 1.4 MeV/u-U^{28+} colliding with atomic hydrogen, is shown as a function of hydrogen density. It is seen that the DE is small for hydrogen densities lower than 10^{20} cm^{-3}, and the EL cross sections show little changes at lower target densities (see, (6.29)).

One can conclude that to describe the charge-changing processes occurring in a dense gas, plasma, or solid-state medium, the standard formulae for effective cross sections, generally speaking, cannot be applied, and the formulae, taking the target density effect into account, should be applied (see, e.g., [42, 182] for detail).

Chapter 7
Interaction of Heavy Ion Beams with Plasmas

Abstract Atomic radiative and collisional characteristics of ion beams with plasmas are required for solving many problems in spectroscopic and particle diagnostics of plasmas (e.g., HIBP method—Heavy-Ion Beam Probe [394], determination of the stopping power of ion beams in plasmas and optimal conditions for stripping of ion beams to obtain the maximum charge states and narrow charge-state distributions etc. As has been mentioned before, interactions of ion beams with plasma particles significantly differ from those with neutral particles in gaseous and solid targets. Moreover, the corresponding interaction cross sections strongly depend on plasma temperature, density, particle abundances and others. In this chapter, atomic processes of fast heavy ions with plasma particles are considered such as radiative and ternary recombinations, dielectronic recombination, ionization by electron impact and others. A Maxwellian distribution function of plasma particles depending on the incident-ion velocity (the so-called *shifted velocity distribution*) is discussed.

7.1 Atomic Processes Between Projectile Ions and Plasma Particles

In plasmas, besides electron loss and capture processes, the following additional charge-changing processes occur:

1. Radiative recombination (RR)—electron capture of free electrons with photon emission:

$$X^{q+} + e^- \rightarrow X^{(q-1)+}(nl) + \hbar\omega, \quad \hbar\omega = E_e + E(nl), \quad E(nl) > 0, \quad (7.1)$$

© Springer International Publishing AG 2018
I. Tolstikhina et al., *Basic Atomic Interactions of Accelerated Heavy Ions in Matter*, Springer Series on Atomic, Optical, and Plasma Physics 98, https://doi.org/10.1007/978-3-319-74992-1_7

where q denotes ion charge, E_e the kinetic energy of a free electron, $E(nl)$ the binding energy of the ion $X^{(q-1)+}$ in the final state with the quantum numbers n and l, and $\hbar\omega$ the photon energy. RR is the inverse process to photoionization and takes place with any energy of a free electron.

2. Dielectronic recombination (DR):

$$X^{q+} + e^- \rightarrow [X^{(q-1)+}]^{**} \rightarrow X^{(q-1)+}(nl) + \hbar\omega, \qquad (7.2)$$

is a two-step process occurring first with a capture of a free electron and simultaneous excitation of an ion inner-shell electron into a doubly excited state, and finally radiatively decaying into nl state of the $X^{(q-1)+}$ ion. Although both processes (7.1) and (7.2) have the same products in the final channel, there is a principal difference between these processes: RR takes place at *arbitrary* free-electron energy, but DR occurs only at specific electron energies E_e satisfying the *resonance* condition:

$$E_e \approx \Delta E - \frac{q^2 Ry}{n^2} < \Delta E, \quad \Delta E > 0, \qquad (7.3)$$

where ΔE is an excitation energy in the $X^{(q-1)+}$ ion. RR and DR processes are schematically shown in Fig. 7.1 as a function of a electron-ion collision energy. Radiative recombination dominates at low electron energies. Resonances associated with DR into excited Rydberg states $n \gg 1$ of $X^{(q-1)+}(nl)$ ion are shown. About RR and DR processes see [395] in detail.

3. Three-body (ternary) recombination (TR)—inverse process to electron-impact ionization:

$$X^{q+} + e^- + e^- \rightarrow X^{(q-1)+} + e^-. \qquad (7.4)$$

Fig. 7.1 Scheme of radiative-recombination and dielectronic-recombination cross sections as a function of a electron-ion collision energy. From [395]

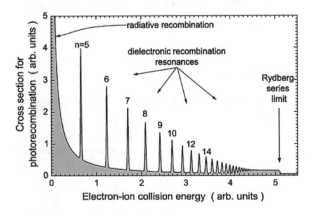

4. Multiple-electron ionization by electron and ion impact:

$$X^{q+} + e^-, \; A^{k+} \rightarrow X^{(q+m)+} + A^{k+} + me^-, \quad m \geq 1. \tag{7.5}$$

where A^{k+} denotes a plasma ion in the charge state k.

5. Ionization of the projectile ion as a result of *ion-ion* electron capture:

$$X^{q+} + A^{k+} \rightarrow X^{(q+1)+} + A^{(k-1)+}. \tag{7.6}$$

Cross sections and rate coefficients of the processes (7.1)–(7.6), averaged over Maxwellian velocity distribution for *isolated* plasma, i.e., without interacting with incoming beams, are considered in many books (see, e.g., [16–25, 396]).

7.2   Radiative Recombination (RR)

RR is one of the main recombination mechanisms of ion beams, passing through plasmas, because it occurs with larger probability than electron capture process.

RR cross sections, (7.1), into final Rydberg states with $n \gg 1$, averaged over orbital quantum numbers l, are usually calculated using the classical Kramers formula [114] or semiempirical formulae [397], introducing an *effective* charge for resulting $X^{(q-1)+}$ ion. For RR cross sections into low-lying levels n, the Kramers cross sections should be multiplied by the quantum-mechanical *Gaunt factor* [398, 399].

Kramers RR cross sections (7.1) into a specific n state and into all n-states of $X^{(q-1)+}$ ion are given with account for free vacancies by:

$$\sigma_{RR}(n, E_e) = 2.105 \times 10^{-22} \, [\text{cm}^2] \frac{Z_{eff}^4 \cdot 13.606^2}{n^3 E_e (E_e + Z_{eff}^2 \cdot 13.606/n^2)}, \tag{7.7}$$

$$\sigma_{RR}^{tot}(n, E_e) = \left(1 - \frac{p}{2n_0^2}\right) \sigma_{RR}(n_0, E_e) + \sum_{n>n_0}^{n_{cut}} \sigma_{RR}(n, E_e), \tag{7.8}$$

where p is the number of equivalent electrons in the ground-state configuration n_0^p of the $X^{(q-1)+}$ ion, and Z_{eff} is an effective charge of the $X^{(q-1)+}$ ion. The kinetic energy of a free electron E_e in (7.7) is given in eV. The parameter n_{cut} is the maximum principal quantum number up to which recombination can be observed and is defined by experimental conditions (high electron density, external electric field, etc.). Usually, the main contribution to the sum in (7.8) is given by recombination into the ground and nearby n-states. At high energies E_e, the cross section $\sigma_{RR}(n, E_e)$ falls off as

$$\sigma_{RR}(n, E_e) \sim q^4/(E_e^2 n^3). \tag{7.9}$$

The effective charge Z_{eff} for many-electron heavy ions can be found from ionization potential I_P of the $X^{(q-1)+}$ ion using the Rydberg formula:

$$Z_{eff} = n_0\sqrt{I_P[\text{eV}]/13.606}, \tag{7.10}$$

where n_0 denotes the principal quantum number of the ground-state configuration. For example, for uranium ions U^{q+}, using the tables for binding energies I_P from [374], the effective charges can be fitted by the formula:

$$Z_{eff}(U^{q+}) = 12.75 + 1.15q^{0.9385}, \tag{7.11}$$

which for $q = 91$ gives $Z_{eff} \approx 92$; here q is the ion charge in the final state.

We note that cross sections for REC processes (4.37) can be also estimated by the Kramers formulae, (7.7)–(7.10) as a function of projectile energy E_P using the relation

$$E_e[\text{eV}] = \frac{1000}{1.8229} E_P[\text{MeV}/u], \tag{7.12}$$

which follows from the collision kinematics (see [291]).

7.3   Dielectronic Recombination (DR)

DR is an important recombination process, playing a key role in formation of plasma ionization balance, slowing down and formation of effective charge of ions in plasmas, as well as in investigation of atomic characteristics of heavy many-electron ions such as the Lamb and isotopic shifts and others (see [395–400]). Moreover, DR spectral lines, the *dielectronic satellites*, are effective tools in plasma diagnostics, i.e., determination of plasma temperature, density, particle abundances etc. (see [21, 402]).

DR cross sections are expressed in terms of radiative and autoionization transition probabilities in the initial, intermediate and final ionic quantum states and exhibit as narrow resonances, intensities I of which drastically increase with the ion charge approximately as q^4 but decreases with the principal quantum number n, i.e., $I \sim q^4/n^3$. The widths of DR resonance lines are proportional to kT_e where T_e denotes a plasma temperature and k the Boltzmann constant. An example of DR cross section (rate) of Li-like Ni^{25+} ions for 2s-2p transition is shown in Fig. 7.2.

We note that experimental and theoretical investigations of DR processes involving heavy *many-electron* ions face severe difficulties related with atomic energy structure of these systems because the number of levels even for the ground state can be very large. Figure 7.3 shows calculated *density of states* (DOS), i.e., a number of levels per energy unit, for W^{19+} ions which are populated due to DR of $W^{20+}(4f^8\ {}^7F_6)$ into $4f^8nl$ excited states. It is seen that DOS is extremely large— about 10^7 levels per 1 eV-energy interval! In these cases the usual calculation methods can not be used and new theoretical approaches have to be found. In the works

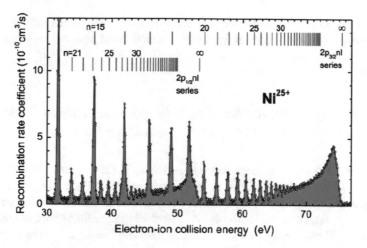

Fig. 7.2 Experimental DR rate coefficient of Li-like Ni^{25+} ions for 2s-2p transition, i.e., cross section averaged over electron-beam velocity distribution function in a storage ring. Principal quantum numbers n for highly excited resonance states are shown. From [403]

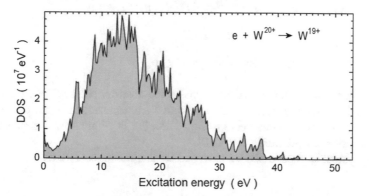

Fig. 7.3 Calculated density of states (DOS) of W^{19+} ions which are populated by DR of W^{20+} comprising all $4f^8 nl$ doubly excited states above the W^{20+} ($4f^8 \, ^7F_6$) ground level with the principal quantum numbers $n < 72$. From [400]

[404–407] a *quantum chaos* method is applied, based on the statistic approach for spectral analysis and eigenvalues of ion quantum states that leads to a better agreement with experiment (see [405]).

7.4 Single- and Multiple-Ionization of Ions by Electron Impact

Single and multiple ionization of ions by electron impact are important processes playing a key role in various physical applications such as plasma kinetics, development of new laser schemes, modeling of laboratory and astrophysical plasmas,

charge-state evolution of atoms and ions exposed to electron beams and others. Here
we consider atomic processes occurring in interactions of projectile ion beams with
plasma free electrons.

7.4.1   Fitting Formulae for Single-Electron Ionization Cross Sections and Rates of Highly Charged Ions

Single-electron ionization cross sections and rates for heavy ions were calculated
in [408–410] using the Coulomb-Born approximation with exchange (CBE). The
fitting parameters for calculated cross sections σ_{nl} and rate coefficients $<\upsilon\sigma_{nl}>$ for
nl states $nl \leq 6h$, including ionization of inner shells and excited states, are presented
in Table 7.1. They were obtained by Least-Square (LS) fitting using the formulae:

$$\sigma_{nl}(u) = p(Ry/I_{nl})^2 \frac{Cu}{(u+1)(u+\varphi)} [\pi a_0^2], \quad u = E/I_{nl} - 1, \quad (7.13)$$

$$<\upsilon\sigma_{nl}>(\beta) = p(Ry/I_{nl})^{3/2} e^{-\beta} G(\beta) [10^{-8} \text{ cm}^3/\text{s}], \quad (7.14)$$

$$G(\beta) = \frac{A\beta}{\beta + \chi}, \quad \beta = I_{nl}/T, \quad (7.15)$$

$$0 \leq u \leq 16, \quad 0.125 \leq \beta \leq 8, \quad (7.16)$$

where p denotes the number of electrons in the electron shell nl^p, I_{nl} the binding
energy of the nl state, E an electron energy and T a plasma temperature. The accuracy
of (7.13)–(7.16) with parameters given in Table 7.1 is within 15% compared to the
CBE numerical calculations [409, 410].

For physical applications the l-averaged ionization cross sections

$$\sigma_n = n^{-2} \sum_{l=0}^{n-1} (2l+1)\sigma_{nl} \quad (7.17)$$

are also required. Using the fitting parameters in Table 7.1 and applying the LS
method, one can obtain the fitting parameters for C, φ and A, χ in the same form
as in (7.13)–(7.16). The results are given in Table 7.2. In this case, the ionization
potentials I_n correspond to the averaged over l values.

For a target ion having more than one electron shell, the total cross section can
contain a structure on top of the direct-ionization (DI) cross section due to indirect
(multistep) ionization mechanisms. One of such mechanisms is excitation of the
inner-shell electrons into autoionizing states followed by *autoionization* decay

$$X^{q+} + e^- \rightarrow [X^{q+}]^{**} + e^- \rightarrow X^{(q+1)+} + 2e^-, \quad (7.18)$$

which is termed as Excitation-Autoionization (EA).

Table 7.1 Fitting parameters for cross sections, (7.13), and rate coefficients, (7.14), for ionization of highly charged ions from nl-states by electron impact with $0.1 \leq u \leq 14$, $0.1 \leq \beta \leq 8$. From [408]

nl state	C	φ	A	χ
1s	7.96	2.70	5.65	0.40
2s	6.69	2.03	6.23	0.52
2P	6.93	1.47	9.05	0.73
3s	6.00	1.59	7.37	0.70
3P	6.24	1.31	9.11	0.82
3d	6.57	1.08	7.76	1.00
4s	5.77	1.43	9.11	0.76
4P	6.00	1.26	11.7	0.86
4d	6.23	1.11	10.8	0.97
4f	7.06	1.00	13.5	1.07
5s	5.66	1.36	7.96	0.79
5P	5.88	1.23	9.13	0.87
5d	6.08	1.12	10.4	0.96
5f	6.26	1.08	11.1	1.00
5g	6.47	1.04	11.9	1.03
6s	5.60	1.32	8.07	0.80
6P	5.82	1.22	9.13	0.88
6d	6.00	1.13	10.2	0.95
6f	6.24	1.07	11.2	1.00
6g	6.33	1.04	11.7	1.03
6h	6.44	1.01	12.2	1.06

Table 7.2 Fitting parameters for the l-averaged cross sections σ_n and rates $< \upsilon \sigma_n$ for ionization of highly charged ions by electron impact, (7.13)–(7.16) and (7.17) with $0.01 \leq u \leq 14$, $0.1 \leq \beta \leq 8$. From [408]

n state	C	φ	A	χ
1	7.96	2.70	5.65	0.40
2	7.82	1.55	8.33	0.68
3	6.44	1.23	10.2	0.90
4	6.30	1.11	10.9	0.97
5	6.24	1.06	11.2	1.01
6	6.21	1.03	11.4	1.03

Figure 7.4 shows the contribution of EA to the total ionization cross section in the case of ionization of U^{16+} and Si^{6+} ions. It is seen that in the case of heavy many-electron ions (U^{16+}), a contribution of EA processes is very large. For medium-heavy ions (Si^{6+}), excitation of inner-shell electrons followed by autoionization is not very significant.

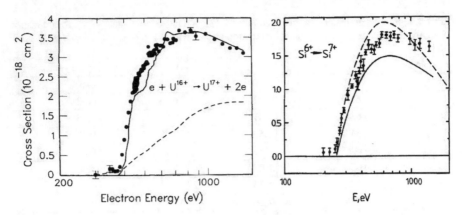

Fig. 7.4 Left: Single-electron ionization cross section of U^{16+} ions by electron impact: dashed curve—direct ionization result with semiempirical Lotz parameters [414] for direct ionization of electrons in the $4f^{14}5s^25p^65d^76s$ metastable configuration. The solid curve is a distorted-wave calculations including excitation-autoionization effects [415]. From [413]. Right: Ionization cross section of Si^{6+} ions: solid curve—distorted-wave calculations [412], dashed curve—(7.13) with inclusion of ionization from the inner 2s and 2p states; circles—experiment [411]. From [410]

The sum of direct ionization (DI) and ionization via excitation-autoionization (EA) cross sections is presented in the form:

$$\sigma = \sigma_{DI} + \sum_j B_j \sigma_{exc}(j), \qquad (7.19)$$

$$B_j = \sum_m \Gamma_{jm} \Big/ \left(\sum_m \Gamma_{jm} + \sum_l A_{jl} \right), \qquad (7.20)$$

where B_j is the *branching-ratio* coefficient of the autoionizing j state of the ion, A and Γ are the radiative and autoionization probabilities, respectively, σ_{exc} is the excitation cross section of inner-shell electrons. The autoionization probability $\gamma \approx 10^{13}-10^{14}\,\mathrm{s}^{-1}$ weakly depends on the ion charge q. For *low-charged* ions $q \le 10$, the radiative probability $A \ll \Gamma$ and therefore $B \approx 1$. In the case of *highly charged* ions, the quantities Γ and A differ significantly depending on the states, and it is necessary to use the general formula (7.20) for the branching ratios that strongly complicates the ionization cross-section calculations (see, e.g., [416, 417]). EA cross sections and rates can be estimated by semiempirical formulae (5.51) or using some fitting approximations [418–420].

Besides DI and EA, there are other important indirect processes leading to ionization but showing resonant characteristics, i.e., they occur only at a definite (resonance) energy of the incident electron. Such resonant processes are multi-step processes and lead to an appearance of narrow resonances in the total ionization cross section. These processes are called Resonant-Excitation-Auto-Double-Ionization

(READI) and Resonant-Excitation-Double-Autoionization (REDA). They were predicted theoretically [421, 422] and confirmed experimentally [423]. Such measurements become possible only by using advanced crossed-beam techniques [425], which, in particular, permit to study narrow features such as resonance and excitation thresholds with extremely high precision to obtain spectroscopic information on multiply excited states [426].

7.4.2 Semiempirical Formulae for Double-Ionization Cross Sections of Heavy Ions by Electron Impact

Double-ionization cross sections $\sigma_2(E)$ of neutral atoms and positive ions by electron impact are required for many applications in plasma physics and accelerator physics, especially if the cross sections are presented in the form of semiempirical formulae which can be used for easy analytical representation of the double-ionization cross sections of heavy positive ions in the modelling of laboratory and astrophysical plasmas.

In [427, 428], the $\sigma_2(E)$-values are considered for incident-electron energies $E < 50 \cdot I_{th}$ where I_{th} is the threshold energy for double-electron ionization and semiempirical formulae are suggested which describe the experimental cross sections within an accuracy of 20–30%. The formulae were obtained on the basis of reliable experimental data and quantum-mechanical calculations.

In [427] simple semiempirical formulae with three fitting parameters are suggested for light positive ions from He- to Ne-like isoelectronic sequences, and for heavy ions of Ar^{q+}, $(q = 1-7)$ and Kr^{q+} $(q = 1-4)$ by taking into account the contribution of direct double-ionization and of inner-shell ionization processes. The suggested formulae can be used for prediction of the double-ionization cross sections of positive ions with the nuclear charge $Z_N \leq 26$ for energies $E < 50 I_{th}$. All fitting parameters are found to be constant for ions within the given isoelectronic sequence. The analysis, made in [427], also provides a method for indirect determination of K-shell ionization cross sections for ions from Be-like to Ne-like sequences. In [427], the fluorescence yields ω_K for a single K-shell vacancy in ions from Li-like to Ne-like sequences with nuclear charges $3 \leq Z_N \leq 26$ are calculated as well.

In Fig. 7.5 (left), experimental $\sigma_2(E)$-values for Ar atoms, $q = 0$, and Ar^{q+} ions, $q = 1-7$, are compared with semiempirical formulae, obtained in [427], as a function of electron energy, and in Fig. 7.5 (right) the same comparison is made for Kr^{2+} ions.

In [428], semiempirical formulae with six fitting parameters are suggested for many-electron heavy ions from Ti up to Bi (nuclear charge from $Z = 22$ to $Z = 83$) for incident electron energies $E < 50 I_{th}$: Ti^{q+}, $q = 1-6$, Fe^{q+}, $q = 1, 3-6$, Ni^{q+}, $q = 1-6$, Ga^{q+}, $q = 1-6$, Kr^{q+}, $q = 1-4$, Mo^{q+}, $q = 1-6$, Pr^{q+}, $q = 1-4$, Sm^{q+}, $q = 1-6$, W^{q+}, $q = 1-6$, Pb^{q+}, $q = 1-9$ and Bi^{q+}, $q = 1-10$, 12. The formulae are derived on the basis of experimental data, mostly performed at an electron-ion crossed-beams set-up in Giessen, Germany, with quantum-mechanical considera-

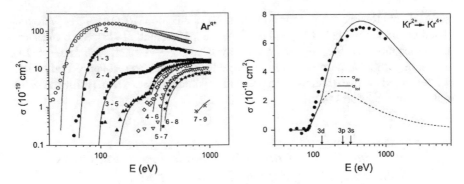

Fig. 7.5 Left: Double-ionization cross sections of Ar atoms, $q = 0$, and Ar^{q+} ions, ($q = 1-7$), by electron impact. Experiment: open circles double-ionization cross sections of neutral Ar [429], other symbols double ionization of Ar^q, ($q = 1-5$) ions [430]. Solid curves double-ionization cross sections calculated by semiempirical formulae. Right: Double-ionization cross sections of Kr^{2+} ions. Experiment: solid circles [425, 429, 430]. Dashed curve direct cross section for simultaneous ejection of two electrons and dotted curve double-ionization cross section, result of semiempirical formulae. Arrows indicate the threshold energies for inner-shell ionization. From [427]

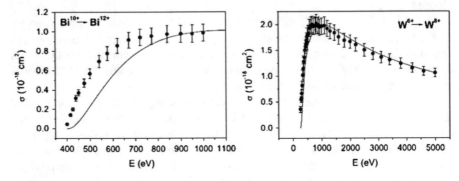

Fig. 7.6 Left Electron-impact double-ionization cross section (in 10^{-18} cm^2) of Bi^{10+} ions. Full circles experiment [431]; solid line result of semiempirical formulae. Right: Electron-impact double-ionization cross section (in 10^{-18} cm^2) of W^{6+} ions. Full circles experiment [432]; solid line result of semiempirical formulae. From [428]

tions. The contribution of direct double ionization of two outer-shell electrons of the ions is also taken into account together with single inner-shell ionization processes followed by autoionization with additional ejection of an electron as was suggested in [427]. The formulae describe well the available experimental double-ionization cross sections within an accuracy of 20–30%. However, for multiple-electron ionization of very heavy ions significant deviations from experiment are found in the low-energy region. These deviations are most probably caused by higher order processes such as inner-shell excitation and subsequent double autoionization (EDA).

In Fig. 7.6, experimental $\sigma_2(E)$-values for W^{6+} and Bi^{10+} ions are compared with semiempirical formulae, obtained in [428], as a function of electron energy.

7.4.3 Semiempirical Formulae for Multiple-Ionization Cross Sections of Atoms and Ions by Electron Impact

At present, cross sections $\sigma_m(E)$ of m-fold ionization by electron impact

$$e^- + A \rightarrow e^- + A^{m+} + me^-, \quad m \geq 2, \quad (7.21)$$

have been measured for most of neutral atoms and a lot of species—negative and positive ions—including highly charged ions (see, e.g., [421–437] and references therein) from threshold I_{th} to electron energies $E \approx 20$ keV.

The most accurate measurements of multiple-electron ionization cross sections by electron impact were performed using the crossed-beam technique with the absolute total uncertainties of the order of a few percents and the smallest investigated cross section range $\sigma_m < 10^{-22}$ cm^2. In contrary to the large volume of the experimental data obtained, a theory of multiple ionization of atoms and ions by electron impact still suffers from a lack of the quantum-mechanical treatment even for double-electron ionization. This is mainly connected with the fact that the independent electron approximation (IEA), used in ion-atom collisions, fails in the description of multi-electron transitions induced by electron impact. Therefore, to predict multiple-ionization cross section behavior and their values, semiempirical formulas are often used.

Semiempirical formulae for m-fold ionization cross sections, $m \geq 2$, of atoms, negative and positive ions by electron impact have been developed in [435, 436] on the basis of experimental data and the Bethe-Born behavior of $\sigma_m(E)$ at high energies in the form:

$$\sigma_m(u) = 10^{-18} \, [\text{cm}^2] \frac{a(m)N^{b(m)}}{(I_{th}/Ry)^2} \left(\frac{u}{u+1}\right)^c \frac{\ln(u+1)}{u+1}, \quad (7.22)$$

$$u = E/I_{th} - 1, \quad m \geq 2, \quad (7.23)$$

where E denotes the incident electron energy, N the *total* number of the projectile electrons, I_{th} the threshold energy, a, b and c fitting parameters and Ry is the Rydberg unit, 1 Ry = 13.606 eV.

The threshold energies I_{th} are given by the sum of ionization potentials:

$$I_{th} = \sum_{q'=q}^{q+m-1} I_{q',q'+1}, \quad (7.24)$$

where $I_{q',q'+1}$ is the single-electron ionization energy from the charge q' to $q'+1$. For single-electron ionization, I_{th} is the first ionization potential, $I_{th} = I_1$. The minimal energy I_{th} for double ionization is the sum of the first and the second ionization potentials of the target, and so on. For example, the minimal energy, required to ionize three electrons in Ar atom is estimated to be: $I_{th} = I(Ar) + I(Ar^+) + I(Ar^{2+}) = 15.8$ eV $+ 27.6$ eV $+ 40.9$ eV $= 84.3$ eV. The threshold ionization energies I_{th} can be estimated using the tables for $I_{q,q+l}$ values for atoms and ions given in [372–375, 438, 439]. The energy required to ionize all N electrons from a *neutral* atom is well described by the statistical Thomas-Fermi formula:

$$I_{th}(m = N) = 16\,N^{7/3}\,[\text{eV}]. \tag{7.25}$$

The exponent c in (7.22) is determined empirically: $c = 1.0$ for neutral atoms and $c = 0.75$ for positive and negative ions. The fitting parameters $a(m)$ and $b(m)$ are smooth functions of the number of ejected electrons m; for $2 \le m \le 10$ they are given in Table 7.3. For ionization of $m > 10$ electrons, the asymptotic values for $a(m)$ and $b(m)$ can be used:

$$a(m) \approx 1350\,m^{-5.7}, \quad b(m) = constant = 2.00, \quad m > 10, \tag{7.26}$$

which are obtained by extrapolation of the $a(m)$ and $b(m)$ values for $m < 10$.

Table 7.3 Fitting parameters $a(m)$ and $b(m)$, (7.22) for m-fold ionization cross sections by electron impact. From [436]

m	$a(m)$	$b(m)$
2	14.0	1.08
3	6.30	1.20
4	0.50	1.73
5	0.14	1.85
6	0.049	1.96
7	0.021	2.00
8	0.0096	2.00
9	0.0049	2.00
10	0.0027	2.00

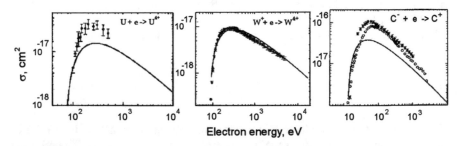

Fig. 7.7 Multiple ionization of neutral atom, positive and negative ions by electrons. Left figure: four-fold ionization of neutral uranium atom U^0: solid circles—experiment [440], solid line—(7.22). Middle: three-fold ionization of W^+ ions: experiment [431], solid line—(7.22). Right: double ionization of C^- ions: experiment—open circles [441], solid circles [436], solid line—(7.22). From [434, 436]

The average accuracy of the semiempirical formula (7.22) is a factor of 2–3 (Fig. 7.7).

7.5 Rate Coefficients of Atomic Processes in Ion-Beam-Plasma Interaction. A Shifted Maxwellian Distribution Function

Plasma constituent particles (atoms, ions, electrons) have a different concentrations depending on plasma temperature and density, therefore, to describe interaction of ions with plasma particles the *rate coefficients* $\kappa = N <v\sigma> [\text{s}^{-1}]$ of the processes are used, i.e., averaged over a Maxwellian velocity distribution of particles where N denotes a particle density in a plasma.

In interaction of ion *beam* with a plasma, a Maxwellian function $F(v, v_p, T)$ depends on the ion beam velocity v_p. Then the quantity $<v\sigma> [\text{cm}^3/\text{s}]$ is given by [26]:

$$<v\sigma> = \int_0^\infty v\sigma(v) F(v, v_p, T) d^3v, \quad \int_0^\infty F(v, v_p, T) dv = 1, \quad (7.27)$$

$$F(v, v_p, T) = \left(\frac{M}{2\pi kT}\right)^{3/2} exp\left(-\frac{M}{2kT}|\mathbf{v} - \mathbf{v}_p|^2\right) \quad (7.28)$$

$$= \left(\frac{M}{2\pi kT}\right)^{1/2} \frac{v}{v_p} \left[exp\left(-\frac{M}{2kT}(v - v_p)^2\right) - exp\left(-\frac{M}{2kT}(v + v_p)^2\right)\right],$$

where $\mathbf{v} = \mathbf{v}_p - \mathbf{v}_{e,i}$ denotes a vector of the incident-ion relative to electron velocity $\mathbf{v}_e$ or ion velocity $\mathbf{v}_i$ in a plasma, M a reduced mass of colliding particles, T an electron or ion temperature, k the Boltzmann constant.

At *low* ion velocity $v_p \to 0$, the function $F(v, v_p, T)$ turns into the 'usual' Maxwellian function $F(v, T)$, and at *low* plasma temperature $2T/M \to 0$ one has:

$$F(v, v_p, T) = \delta(v - v_p), \quad 2T/M \to 0. \quad (7.29)$$

and

$$< v\sigma > \approx v_p\sigma(v_p), \quad v_p \gg v_{th} = 1.13\sqrt{\frac{2T}{m}}. \quad (7.30)$$

Here v_{th} is the thermal electron velocity which in equilibrium plasmas is much higher than the thermal ion velocity due to the difference in masses. Therefore, rate coefficients of atomic interaction processes of heavy-ion beams passing through cold plasmas are simply defined by the *product* $N v\sigma_p$ of the ion velocity and cross section which leads to two important conclusions:

1. Rate coefficients of fast ions in cold plasmas ($v_p \gg v_{th}$) are independent on the plasma temperature and plasma particle distribution over velocities,

2. In the case of low ion-beam velocities, $v_p \lesssim v_{th}$, the rate coefficients are defined by the general formulae (7.27) and (7.28), i.e., using a shifted Maxwellian function depending on the ion-beam velocity v_p.

7.6 The Case of Fast Ion Beams Penetrating Cold Plasmas

The case $v_p \gtrsim v_{th}$ is usually achieved in practice for which the rate coefficient is given by:

$$\kappa = N v_r \sigma(v_r) [\text{s}^{-1}], \quad v_r \approx \sqrt{v_p^2 + v_{th}^2}, \tag{7.31}$$

where v_r is a *relative* velocity and N is a density of plasma particles.

Three-body recombination (TR), (7.4), constitute a special case because its rate does not change linearly on electron density like in (7.31) but is proportional to N_e^2. The TR rate κ_{TR} can be calculated in the classical approximation [443]:

$$\kappa_{TR} = \frac{2^5 \pi^2 e^{10} N_e^3 q^3}{m^5 v_r^9} \approx 2.9 \times 10^{-31} [\text{s}^{-1}] \frac{N_e^2 [\text{cm}^{-3}] q^3}{v_r^9 [\text{a.u.}]}, \tag{7.32}$$

where q denotes the incident ion charge and v_r a relative velocity in atomic units, 1 a.u. $\approx 2.2 \times 10^8$ cm/s.

A comparison between radiative-recombination (7.1) and TR rates gives an estimate [75]:

$$\frac{\kappa_{RR}}{\kappa_{TR}} \approx 1.6 \times 10^{17} \frac{q v_r^6 [\text{a.u.}]}{N_e [\text{cm}^{-3}]}, \tag{7.33}$$

from which it is seen that the TR rate is small and close to the RR rate only at very high electron density N_e.

A comparison of electron-capture (EC) and radiative recombination (RR) recombination rates gives:

$$\frac{\kappa_{EC}}{\kappa_{RR}} \approx 10^7 \frac{N_{at}}{N_e} \frac{q Z_T^5}{v_r^8 [\text{a.u.}]}, \tag{7.34}$$

where N_{at} and N_e denote densities of neutral atoms and free electrons in a plasma, respectively, Z_T the atomic number of neutral atoms. It is obvious that this ratio strongly depends on the relative velocity v_r and on the ratio N_{at}/N_e.

Concentrations (abundances) of atomic particles in plasmas depend on plasma temperature and density. As an example, the abundances of atoms and ions in H and He plasmas, calculated by solving the balance equations in the *coronal limit* [442], are shown in Fig. 7.8 as a function of electron temperature T_e; in the coronal limit, abundances depend only on the electron temperature but not on plasma density.

In a cold hydrogen plasma with $T_e = 1$ eV (left figure), the ratio $N_{at}/N_e = N_H/N_{H^+} \approx 30$, and for ions with the charge $q = 20$ and energy 1.5 MeV/u ($v_r = 7.7$ a.u.) the rate ratio (7.34) is about 10^3. It means that in a cold plasma, electron

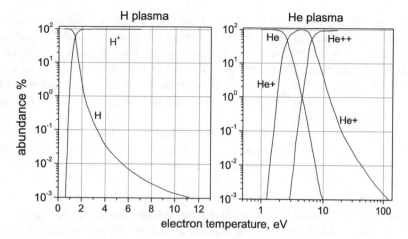

Fig. 7.8 Abundances of neutrals and positive ions in H (left) and He (right) plasmas as a function of electron temperature—calculations in the coronal limit [442]

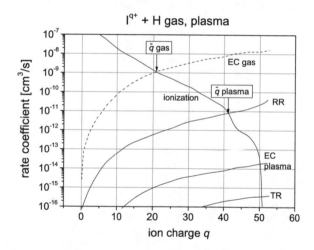

Fig. 7.9 Calculated ionization and recombination rate coefficients for 1.5-MeV/u I^{q+} ions passing through hydrogen gas and plasma with parameters $T_e = 10\,\text{eV}$ and $N_e = 10^{17}\,\text{cm}^{-3}$. EC gas (dashed line) and EC plasma (solid line) correspond to electron-capture rates of bound electrons in a gas and plasma, respectively. Curve *ionization* is ionization rate which is nearly the same for a gas and plasma targets at $q > 5$. The curves RR and TR are radiative-recombination and three-body recombination rates, respectively. DR processes are not accounted for. The arrows show the mean charges of iodine ions created in hydrogen gas and plasma. The rate data are from [75]

capture rate of 1.5-MeV/u beam ions with $q = 20$ on plasma neutral atoms is much larger than the radiative recombination rate.

In an almost fully ionized H-plasma with $T_e = 10$ eV, the ratio $N_{at}/N_e \approx 10^{-5}$ and $\kappa_{EC}/\kappa_{RR} \approx 10^{-3}$, i.e., in strongly ionized plasmas, RR is the main recombination process. This is clearly seen from Fig. 7.9, where the rates of ionization and recombination processes of I^{q+} ions in hydrogen cold gas and plasma are shown. The rates are calculated in [75] with the following parameters: for iodine ion charges $0 \leq q < 60$, energy 1.5 MeV/u and for H plasma with $T_e = 10\,\text{eV}$ and $N_e = 10^{17}\,\text{cm}^{-3}$. Dielectronic recombination are not included in Fig. 7.9. For ions with $q > 5$, ionization rates are nearly the same for hydrogen cold gas and plasma. As seen from Fig. 7.9, the RR rate is much larger than EC rate, and TR rate is much smaller than EC and RR rates in a plasma.

From Fig. 7.9 it is possible to estimate the average mean charges $\bar{q}$ of iodine ions in a hydrogen cold gas and the plasma: $\bar{q}_{gas} \approx 20$ and $\bar{q}_{plasma} \approx 40$. Therefore, plasmas are more effective stripper for fast heavy ions compared to a cold-gas target.

Finally we note that the role of DR processes on recombination rates of fast ions moving through a cold plasma is very important since DR cross sections are proportional to q^4 but has a resonance character: DR contribution can be large or small depending on matching between ion velocity and the energy profiles having a width $\sim kT_e$. The problem of DR contribution to the recombination rates is discussed in [75] but still requires a further detailed consideration, especially in the case of heavy many-electron projectile ions.

Chapter 8
Multiple-Electron Ionization of Atoms by Fast Heavy Ions

Abstract Single- and multiple-electron ionization of atoms and molecules, induced by fast highly charged ions, constitute an important issue in accelerator physics (the vacuum conditions) and in different applications (cancer therapy). Multiple ionization of heavy target atoms can strongly (up to 50%) contribute to the total ionization cross sections similar to *projectile* ionization (loss) by target atoms, Sect. 1.4. In this Chapter, various theoretical approaches for multiple-electron probabilities and cross sections for ionization of neutral atoms are considered and compared with available experimental data.

8.1 Main Properties of Single- and Multiple-Ionization of Atoms by Ions

Ionization of media atoms and molecules by heavy accelerated ions is an important process in atomic physics and its applications. The target-ionization cross sections are quite large because they are proportional approximately to the square of the incident projectile charge, q^2, significantly contributing to several collisional characteristics such as the energy loss, straggling, stopping power, mean charge of ion beams etc. A special case is ionization of residual-gas atoms and molecules by ions in powerful accelerators leading to the *dynamic vacuum effects* at very high ion-beam densities which strongly influences the beam lifetimes of accelerated ions [224, 379]. On the other hand, *multiple* ionization (MI) of target atoms greatly contributes to the total target ionization cross sections (up to about 50%) and, therefore, is an interesting issue for understanding multiple-electron nature of atomic processes, related to electron-electron correlation effects.

As has been mentioned in Sect. 1.4, transfer ionization (TI) is the most general reaction occurring in ion-atom collisions:

$$X^{q+} + A \rightarrow X^{q'+} + A^{m+} + (q' - q + m)e^-. \tag{8.1}$$

© Springer International Publishing AG 2018
I. Tolstikhina et al., *Basic Atomic Interactions of Accelerated Heavy Ions in Matter*, Springer Series on Atomic, Optical, and Plasma Physics 98,
https://doi.org/10.1007/978-3-319-74992-1_8

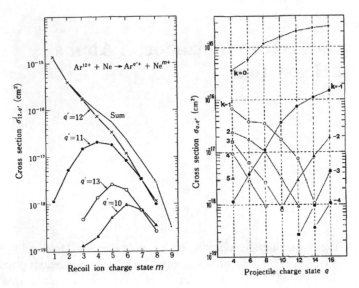

Fig. 8.1   Left: Experimental partial cross sections $\sigma^i_{qq'}$ for recoil Ne^{m+}-ion production in Ar^{12+} + Ne → $\mathrm{Ar}^{q'+}$ + Ne^{m+} collisions at 1.05 MeV/u. The curve with $q' = 12$ corresponds to pure target ionization (8.2), other curves are transfer ionization reactions (8.1), and the curve Sum denotes a sum of all partial cross sections. Right: Total (summed over all recoil charges m) experimental charge-changing cross sections in Ar^{q+} + Ne → $\mathrm{Ar}^{q'+}$ + Ne^{m+} + $(q' - q + m)e^-$ collisions at 1.05 MeV/u. The curve with $k = q' - q = 0$ represents the pure target multiple-electron ionization cross sections. Curves with $k = 1,\dots, 5$ show cross sections for single- to five-fold electron-loss cross sections and curves with $k = -1, \dots, -4$ for single- to four-fold electron-capture cross sections in transfer ionization. From [444]

In some cases, it is experimentally possible to measure partial TI and electron capture (EC) cross sections, as shown in Fig. 8.1, left and right. Cross sections with $q = q'$, i.e., with no change of the projectile charge, correspond to the so-called *pure* target multiple-electron ionization:

$$X^{q+} + A \rightarrow X^{q+} + A^{m+} + me^-, \quad m \geq 1. \tag{8.2}$$

Reaction (8.2) is a more complicated one compared to ionization by electron impact because of a strong influence of the complex atomic structure of the projectile ion.

We note that the data on the experimental partial cross sections, such as displayed in Fig. 8.1, is rather scarce (see [33, 445–447] and references therein) because such measurements should be carried out utilizing the *coincidence technique* to get information on the recoil ions correlated with the projectile charge states [448]. Experimentally and theoretically, pure target ionization processes (8.2) are investigated in more detail.

Experimental loss, capture and pure target ionization cross sections in Ar^{q+} + Ne → $\mathrm{Ar}^{q'+}$ + Ne^{m+} collisions at 1.05 MeV/u, summed over all recoil charges i, are shown in Fig. 8.1, right, as functions of ion charge and a charge-state change

$k = q' - q$. As the projectile charge q increases, the total cross section for pure target ionization ($k = 0$) increase while single- ($k = 1$) and multiple-electron ($k >$ 1) transfer ionization cross sections decrease. The electron-capture cross sections ($k < 0$) increase as the projectile charge increases, and the largest contribution is given by single-electron capture ($k = -1$).

Ionization of atoms by highly charged ions has been one of the main topics of atomic physics investigations in the 1980s and a few decades later. Experimental data on single- and multiple-electron ionization cross sections of atoms and molecules have mainly been obtained at BEVALAC, Berkeley, UNILAC, Darmstadt and RIKEN, Saitama accelerators for gaseous targets He, Ne, Ar, Kr and Xe and projectile ions from protons to bare uranium in 1–420 MeV/u energy range (see, e.g., [1, 6, 33, 449–484]). An overall accuracy of measured MI cross sections is quite large, of 30–50%, due to experimental difficulties.

Some examples of pure target ionization are given in Fig. 8.2 for Ar and Xe atoms, showing large cross sections and little effects from the target atomic structure. In these figures, electron-capture cross sections are also displayed, which become comparable to the pure ionization cross sections with the number of ejected electrons $m = 14$ and 25, for Ar and Xe, respectively. In the case of Xe target, simultaneous ionization of $m = 33$ electrons (!) in one collision has been observed.

Due to the recent fast developments of high-power accelerators and new applications, an interest in investigation of target multiple-ionization processes has been greatly increased again. For example, lifetimes and energy losses of accelerated ion beams strongly depend on their interaction with residual-gas atoms and molecules [224, 379]. These data are of a special interest for a recently started International project FAIR (Facility for Antiproton and Ion Research) [51], Darmstadt, Germany, and Russian project NICA (Nuclotron-based Ion Collider fAcility) in Dubna.

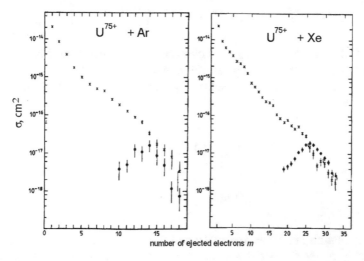

Fig. 8.2 Experimental pure target ionization of Ar and Xe atoms (crosses) and electron capture (circles) cross sections in U^{75+} + Ar (left) and Xe (right) collisions at 15.5 MeV/u. From [449]

Another example of the importance of target ionization by accelerated ions is related to developments of heavy-ion cancer therapy, where the influence of secondary electrons in a body, caused by multiple ionization of surrounding atoms by MeV/u-projectiles, can create more damage than the projectiles themselves (see, e.g., [13]).

8.2   Cross-Section Calculations and Comparison with Experiment

Multiple-ionization cross sections and probabilities of neutral atoms by fast heavy ions are calculated using several basic treatments: the independent-particle model (IPM) [456], the n-particle classical trajectory Monte Carlo method (nCTMC) [240, 366], a classical energy-deposition model [369] and a quantum statistical time-dependent meanfield theory [457–461]. Recently a method based on a combination of classical and quantum approaches is suggested [462, 463] for using in a wide energy range including relativistic domain. In general, multi-particle processes constitute a basic problem for a theoretical investigations, in particular, for collisions of heavy projectiles with many-electron targets.

Independent Particle Model (IPM) is the basic approach used in ion-atom collisions, where target electrons are treated independently from each other and electron-electron correlations are usually neglected (see, e.g., [279, 456, 464]). In the IPM approach, the probability P_m for ejection of m electrons from the target atomic shell with N electrons is described by the *statistical binomial* distribution

$$P_m = C_N^m p_s^m (1 - p_s)^{N-m}, \quad C_N^m = \binom{N}{m} = \frac{N!}{m!(N-m)!}, \tag{8.3}$$

where C_N^m denotes the binomial coefficient and p_s a *single-electron removal* probability depending on the impact parameter b and ion velocity v.

For a target atom with many electron shells, the probability P_m can be written in the form [465]:

$$P_m(b, v) = \sum_\gamma \prod_{i=1}^T C_N^m p_i^{m_i} (1 - p_i)^{N_i - m_i}, \tag{8.4}$$

$$\sum_{i=1}^T m_i = m, \quad \sum_{i=1}^T N_i = N_{tot}, \tag{8.5}$$

where summation on γ is made over all target electron shells T, v denotes a projectile-ion velocity, N_i the number of equivalent electrons in the i-th shell and N_{tot} the total number of the target electrons.

The multiple-electron $\sigma_m(v)$ and total $\sigma_{tot}(v)$ ionization cross sections are then given by

$$\sigma_m(v) = 2\pi \int_0^\infty b\,db\ P_m(b, v), \quad \sigma_{tot}(v) = \sum_{m=1}^{N_{tot}} \sigma_m(v). \tag{8.6}$$

In the IPM approach, single-electron ionization probabilities p_{nl} of the nl target shell are usually calculated in the Born Approximation (Plane-Wave Born Approximation—PWBA) [466–468], and semiclassical approximations such as the deposited energy model [369] and CTMC approach [366]. Here n and l are the principal and orbital quantum numbers, respectively. The IPM approach has some serious disadvantages, e.g., it neglects the change of the target ionization potential during collision that leads to a wrong dependence of ionization cross section on m, especially, for large m-values. Also, the use of IPM for MI ionization of *molecules* is questionable because it requires to make additional approximations which ignore the molecular structure [469].

CTMC approach was developed in [240, 366, 470] and [471] (see also Sect. 4.2). In the CTMC-IPM approach the motion of the projectile and target electrons are described classically. The CTMC is based on the numerical solution of a system of the Hamilton classical-motion equations for all projectile and target electrons N using a large number of impact parameters (about a few thousand) for particle trajectories. The system consists of $6(N + 2)$ nonlinear first-order equations (about a few hundred) in partial derivatives and is solved numerically for the coordinates and momenta for all N electrons and two nuclei in Cartesian coordinates. The use of the CTMC method is quite complicated, because many electrons and atomic trajectories should be taken into account to get enough statistics for the calculated cross sections.

Figure 8.3 shows multiple-ionization cross sections of Ar atoms by U^{90+} at 120 MeV/u, calculated in the CTMC approach, with and without including Auger autoionization processes, in comparison with experimental data. We note that an inclu-

Fig. 8.3 MI cross sections of Ar atoms by U^{90+} ions at 120 MeV/u. Experiment: open circles. Theory: dashed curve—CTMC result without autoionization included and solid curve with autoionization included. From [452]

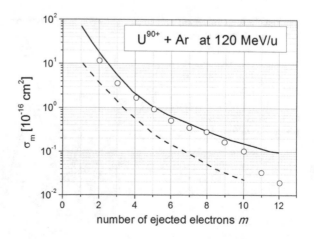

sion of Auger processes in MI calculations is a very important (but quite complicated) problem which will be shortly discussed below.

Energy-deposition (ED) method is the classical model suggested by Bohr [154], based on an assumption that the projectile kinetic energy T_b, transferred to the target electrons, should exceed the first ionization potential I_1 of the target atom, $T_b > I_1$, so that the target atom can eject one or more electrons.

The energy-deposition (ED) model was further developed by Cocke [369], using the statistical model of Russek and Meli [371] for m-electron ionization probability $P_m(b)$, and explained experimental data on multiple-ionization of atoms by fast highly charged ions. The ionization probability $P_m(b)$ in the Russek-Meli model depends on the energy T_b, transferred to the target which is assumed to be statistically distributed among all electrons of the system (see also Sect. 6.2.2).

The ED model [369] has been later developed in [368] for calculation of multiple-ionization cross sections of *ions* by atoms. As it has been mentioned before (Sect. 6.2.2), these two cases of ion-atom collisions are not identical: ionization of *atoms* by ions occurs upon interaction of an atomic electron with a long-range Coulomb field of the projectile, whereas ionization of a projectile *ion* by an atom is due to electron interaction with a field of the neutral target which is close to the Coulomb one at small distances and is exponentially small at large distances from atomic nucleus. The ED approach with the Russek-Meli probabilities provides the unitarity of the ionization probability $P_m(b, v)$ in the form:

$$\sum_m P_m(b, v) = 1 \qquad (8.7)$$

for all ion velocities v and all impact parameters b.

An example of multiple-ionization probabilities $P_m(b)$ and the deposited energy T_b, calculated in the ED model, is shown in Fig. 8.4 for collisions of 1 MeV/u-X^{9+} ions with Ne atoms as a function of impact parameter.

Fig. 8.4 Multiple (m-fold) ionization probability $P_m(b)$, $m = 1$–7, and energy T_b, deposited to a Ne atom, bombarded by 1-MeV/u ions with a charge $q = 9$, calculated by the ED model as a function of impact parameter b. From [369]

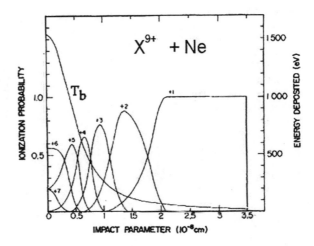

Time-dependent mean-field theory (TDMF) has been developed in [457–461] and is based on the description of time-dependent many-particle systems in terms of the quantum-statistical method. Time dependence of a system density is calculated using the classical Vlasov equation in a phase space. In the TDMF theory, a time evolution of the target electrons as a function of a nuclear motion for a fixed impact parameter is similar to that obtained in the time-dependent Hartree-Fock method. A typical example of TDMF results for production of Ne^{m+} recoil ions in collision with uranium ions is shown in Fig. 8.5 in comparison with experimental data.

Geometrical model is a simple semiempirical method to obtain a single-electron ionization probability $P_s(b, \upsilon)$ in the IPM approach, (8.3), in the form [473]:

$$p_s(b, \upsilon) = p_s(0, \upsilon) \cdot \exp(-b/b_{max}), \tag{8.8}$$

where $p_s(0, \upsilon)$ and b_{max} are the approximation parameters found from experiment. The use of (8.8) gives quite a good description of experimental MI cross sections for the *small* number m of ejected electrons (see, e.g., [444, 474]), when the pure ionization processes are due to direct ionization of the target outermost electrons. For multiple-electron ionization, $m \gg 1$, the use of this approach leads to *underestimated* cross-section values, that is most probably due to the Auger effect which is not accounted in the IPM: ejection of inner-shell electrons leads to electron cascades resulting in enhancement of the target ionization cross sections (see [444]).

A *geometrical model* (GM) has been developed in [475–480] and is used for calculation of the single-electron ionization probability $p_{nl}(0)$ at zero impact parameter $b = 0$. The GM model is based on the semiclassical approximation in which the probability $P_{nl}(0)$ has the from [477]:

$$p_{nl}(0) = 1 - \int_{b_e}^{\infty} R_{nl}^2(r) \left[1 - (b_e/r)^2\right]^{1/2} r^2 dr. \tag{8.9}$$

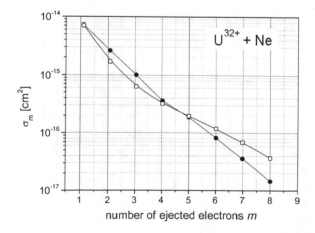

Fig. 8.5 The m-fold electron ionization cross sections of Ne atoms in collisions with 1.4 MeV/u-U^{32+} ions X^{9+} as a function of m. Theory: curve with open circles—the TDMF model. Experimental data—curve with solid circles—are normalized to theoretical data at $m = 1$. From [472]

Here b_e is the maximum radius of the cylinder along the projectile trajectory, inside which the target electron gains the energy higher than the binding energy I_{nl} due the interaction with the Coulomb field of the projectile ion. Equation (8.9) the ionization probability is averaged over the projections of the angular momentum l.

The radius b_e is given by:

$$b_e = \frac{2q}{\upsilon \upsilon_{nl}} V[G(V)]^{1/2}, \quad V = \upsilon/\upsilon_{nl}, \quad \upsilon_{nl} = (2I_{nl})^{1/2}, \tag{8.10}$$

where q is the projectile ion charge, υ and υ_{nl} denote the ion and the atomic orbital velocities, respectively, in atomic units. The radial wave-functions R_{nl} are normalized as

$$\int_0^\infty R_{nl}^2(r)\, r^2 dr = 1. \tag{8.11}$$

Therefore, if the radial wave-functions are known, the single-ionization probability $p_{nl}(0)$ can be easily calculated.

The dimensionless function $G(V)$ is the Gerjuoy-Vriens-Garcia function obtained in the classical BEA (binary encounter) approximation (see [485]). The function $V[G(V)]^{1/2}$ is displayed in Fig. 8.6 and has the following asymptotic behavior:

$$V G(V)^{1/2} \rightarrow \begin{cases} (5/3)^{1/2} \approx 1.291, & V \geq 10, \\ \frac{2}{\sqrt{15}} V^3, & V \ll 1. \end{cases} \tag{8.12}$$

Figure 8.6 shows a strong dependence of $V[G(V)]^{1/2}$ on ratio $V = \upsilon/\upsilon_{nl}$. Therefore, the probabilities $p_{nl}(0)$ are also very sensitive to the ratio V in a whole projectile velocity range υ.

At high-velocity regime $V \gg 1$, it follows that $V[G(V)]^{1/2} \approx (5/3)^{1/2}$, $b_e \approx \frac{2q}{\upsilon \upsilon_{nl}} (5/3)^{1/2}$ and one arrives to a well-known Rutherford formula for the

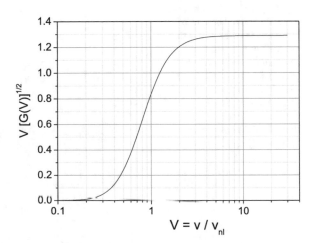

Fig. 8.6 The function $V[G(V)]^{1/2}$ versus the ratio $V = \upsilon/\upsilon_{nl}$ where $G(V)$ is Gerjuoy-Vriens-Garcia function. From [485]

total ionization cross section (see [34, 486]):

$$\sigma_R = b_e^2 \, [\pi a_0^2] = \frac{20}{3} \left(\frac{q}{\upsilon \upsilon_{nl}}\right)^2 \, [\pi a_0^2], \tag{8.13}$$

In [478], the $p_{nl}(0)$ values for $nl = 1s, 2s, \ldots, 4p, 4d$ electron shells are obtained in a close analytical form using H-like radial wave-functions with effective charge accounting the electron screening. For the closed atomic shells, these formulae give a good result for ionization probabilities $p_n(0)$ averaged over l, compared to experimental data as seen in Fig. 8.7.

Normalized Exponential Method (NEM) is introduced and used in [462] and [463]. Recently, a simple method was presented in [463] to estimate multiple-ionization cross sections of atoms by fast ions in a wide energy range $E = 1\,\mathrm{MeV/u}{-}10\,\mathrm{GeV/u}$. The NEM model, described in [463], uses a combination of semiclassical and quantum-mechanical approaches and includes the following steps:

1. Calculation of single-ionization probabilities $p_{nl}(b)$ for the shell nl^N in the form (8.8):

$$p_{nl}(b) = p_{nl}(0) \cdot \exp(-\alpha_{nl}b), \quad \alpha_{nl} = \left[\frac{2p_{nl}(0)}{\sigma_{nl}(\upsilon)/(N \cdot \pi a_0^2)}\right]^{1/2}, \tag{8.14}$$

$$\sigma_{nl}(\upsilon) = 2\pi \int_0^\infty b\,db \; p_{nl}(b) = 2\pi \, \frac{p_{nl}(0)}{\alpha_{nl}^2}, \tag{8.15}$$

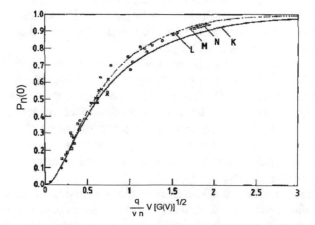

Fig. 8.7 Ionization probability $p_n(0)$ per electron at zero impact parameter for closed K, L, M and N target electron shells as a function of $q/(\upsilon n) \cdot V[G(V)]^{1/2}$ parameter, where n is the principal quantum number of the shell. Symbols—experiment: crosses [481] and open circles [482]—from X-ray data; open squares—Auger data [483]; full circles [484] and [478]. Theory—curves: calculations in the GM using the H-like radial wave functions [478]. From [478]

with $p_{nl}(0)$ is obtained in the GM model [478], single-ionization cross section $\sigma_{nl}(\upsilon)$ of the nl-electron is calculated in the relativistic Born approximation by the RICODE-M program [113], and N is the number of electrons in the nl-state.

2. Calculation of multiple-ionization probabilities $P_m(b, \upsilon)$ for in the IPM approach, (8.4)–(8.6), using a MIT program (Multiple Ionization Transitions) described in [462].

For practical applications the function $V[G(V)]^{1/2}$ in the interval $0.02 < V \leq 10$ was approximated within 10% accuracy by the eight-order polynomial in the form [463]:

$$\log_{10}\left[V[G(V)]^{1/2}\right] = \sum_{k=0}^{8} A_k \cdot (\log_{10} V)^k, \tag{8.16}$$

where the approximation parameters are: $A_0 = -0.09$, $A_1 = 1.17$, $A_2 = -1.94525$, $A_3 = 0.77271$, $A_4 = 0.50108$, $A_5 = -0.35763$, $A_6 = -0.01945$, $A_7 = 0.04806$, $A_8 = -0.0078$.

NEM calculations of MI cross sections for Ne and Ar atoms by Ar^{8+}, Fe^{20+}, Au^{24+}, Bi^{67+} and U^{90+} ions are presented in Figs. 8.8 and 8.9 in comparison with experimental data and CTMC calculations.

Single- and m-fold ionization cross sections of Ne and Ar atoms by highly charged ions are shown in Fig. 8.8 as a function of m. NEM results (open circles) agree within a factor of 2 with experimental data (solid circles) for $m \leq 8$. An exception is the case of $Ne + U^{90+}$ collisions where experimental data are much higher than the NEM calculations that is most probably due to experimental error. The overall agreement of a factor of 2 is because the single-electron cross sections $\sigma_{nl}(\upsilon)$ are calculated in the Born approximation which at some energies, considered here, overestimates experimental data. The use of more sophisticated theoretical result or experimental data for $\sigma_{nl}(\upsilon)$ would give a better agreement.

CTMC results [451, 452] for multiple-ionization of Ar atoms by Fe^{20+} and Au^{24+} ions are also shown in Fig. 8.8 with Auger autoionization included. For Fe^{20+} projectiles an agreement of CTMC calculations with experiment is quite poor but better for Au^{24+} projectiles. In general, CTMC calculations show that Auger cascades play an important role in MI processes, especially, for production of recoil target ions with high charge states. Similar conclusion about importance of Auger processes was also made in [487] and [465], where MI ionization of Ne and Ar atoms by protons were calculated by IPM model.

We note that a number of calculations of multiple-ionization cross sections of heavy atoms with account for Auger decay are quite scarce because of limited experimental data on relative abundances, used in the calculations, for target recoil ions formed due to the sudden inner-shell vacancies (see, e.g., [488]). For this reason, calculations of MI cross sections by IPM approach with Auger processes accounted for are available mainly for Ne atoms with $m \leq 4$ and Ar with $m \leq 7$. In the work [463], $\sigma_m(\upsilon)$ cross sections are calculated for all m: $m \leq 10$ for Ne and $m \leq 18$ for Ar but without accounting for Auger decay. To include Auger processes in calculations

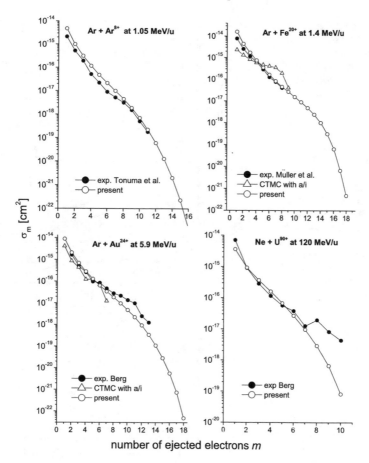

Fig. 8.8 σ_m cross sections as a function of m for ionization of Ne and Ar atoms by highly charged ions. Experiment—solid circles: $Ar + Ar^{8+}$ [444], $Ar + Fe^{20+}$ [451], $Ar + Au^{24+}$ [452], $Ne + U^{90+}$ [452]. Theory: triangles—CTMC calculations with autoionization included for $Ar + Fe^{20+}$ [451] and $Ar + Au^{24+}$ [452] collisions; open circles—NEM calculations. From [463]

of MI cross sections is a quite complicated problem, the solution of which needs a special consideration and is not considered here.

Energy dependence of the total and MI cross sections for ionization of atoms with ejection up to $m = 5$ electrons are shown in Fig. 8.9. Symbols denote experimental data [455] for MI (solid circles) and total cross sections (open circles) of $Ne + Au^{24+}$ and for $Ar + Bi^{67+}$ collisions. Calculated MI and total ionization cross sections [463] are shown by thin and thick curves, respectively. The total cross sections are proportional to the square of the projectile charge: $\sigma_{tot}(v) \sim q^2$.

Experimental data for $Ar + U^{90+}$ collisions in Fig. 8.9 are also compared with a single-electron Born cross section σ_{Born} calculated in the relativistic approximation by the RICODE-M program [365] with account for ionization of all target electrons:

Fig. 8.9 The m-fold and total σ_{tot} cross sections for ionization of Ne and Ar atoms colliding with Au^{24+}, Bi^{67+} and U^{90+} ions as a function of ion energy. Experiment: solid circles—Ne + Au^{24+} at 5.9 MeV/u, $m = 1$–5; Ar + Bi^{67+} at 300 MeV/u, $m = 1$–5; Ar + U^{90+} at 120 MeV/u, $m = 1$–5; open circles—the total cross sections. All experimental data are from [455]. Theory, NEM calculations: thin solid curves—σ_m, $m = 1$–5, and thick curves—σ_{tot}. In the lower figure, dashed curve: σ_{Born}—relativistic Born approximation for one-electron ionization cross section, RICODE-M program, (8.17). From [463]

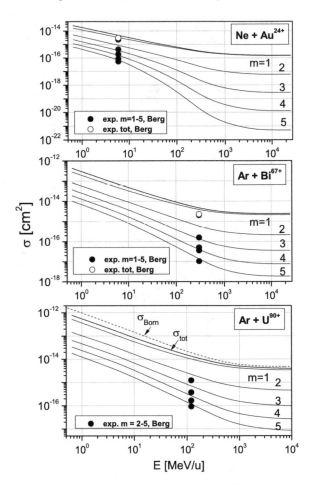

$$\sigma_{Born} = \sum_{nl} \sigma_{nl}(\upsilon). \tag{8.17}$$

The Born cross section $\sigma_{Born}(\upsilon)$ and the NEM single-electron $\sigma_1(\upsilon)$ cross section coincide at high energies but at intermediate energies $E \sim 1$–10 MeV/u, $\sigma_1(\upsilon)$ is smaller than $\sigma_{Born}(\upsilon)$ by a factor of 2 because in calculation of $\sigma_1(\upsilon)$ a normalized ionization probability $p_{nl}(b)$ is used. At relativistic energies $E \geq 1$ GeV/u, Born cross sections turn to semiconstant values because of the influence of the relativistic interactions between colliding particles.

Relative contributions of MI cross sections to the total ones for ionization of Ne and Ar atoms by highly charged ions are shown in Table 8.1. At intermediate ion energies, these contributions are large and decrease with the energy increasing but remain still high (15–20%) even at relativistic energies 1–10 GeV/u (U^{90+} projectile).

Table 8.1 Relative contributions (%) of MI processes to the total target ionization cross sections, $\sum_m \sigma_{m\geq 2}/\sigma_{tot}$, as a function of ion energy. From [463]

Reaction	1 MeV/u	10 MeV/u	10^2 MeV/u	10^3 MeV/u	10^4 MeV/u
$Ne + Ar^{10+}$	31.4	16.9	3.65	0.96	0.75
$Ne + Au^{24+}$	35.6	28.7	14.1	5.0	4.0
$Ne + U^{90+}$	36.3	35.8	30.3	23.2	21.4
$Ar + Au^{24+}$	34.3	23.3	9.3	3.1	2.4
$Ar + U^{28+}$	33.9	24.8	11.2	4.0	3.2
$Ar + Bi^{67+}$	34.6	33.3	21.5	12.5	10.8
$Ar + U^{90+}$	33.6	33.5	25.6	16.7	14.9

The results, discussed in this Section, show that experimental target MI cross sections can be reproduced within a factor of 2–3. A contribution of multiple-ionization cross sections to the total ones can be large $\sim$35% at intermediate energies and decreases with E increasing. However, even at relativistic energies, $E \sim 1$–10 GeV/u, the contribution can be rather large $\sim$20% for heavy target atoms and highly charged projectile ions.

Chapter 9
Some Applications of Charge-Changing Cross Sections and Charge-State Fractions

Abstract In this chapter, some applications of results, obtained in the previous chapters for charge-changing cross sections and equilibrium charge-state fractions, are discussed for ion-beam lifetimes, inverse population in a laser plasma, detection of super-heavy elements, and material modifications.

9.1 Lifetimes of the Heavy-Ion Beams in Accelerator Devices

One of the most important aspects of accelerator physics is the lifetimes of ion-beams in accelerators or storage rings, which principally depend on the atomic charge-changing cross sections of the ion beam interacting with the residual-gas components—atoms, molecules and ions. The lifetime τ for heavy-ion beams injected into an accelerator is defined by

$$I(t) = I_0 \exp(-t/\tau), \tag{9.1}$$

where I_0 denotes the initial intensity of the injected ion beam, and $I(t)$ is its time evolution.

9.1.1 Dependence of Ion-Beam Lifetimes on the Projectile and Target Characteristics

The lifetime τ depends on the so-called *vacuum conditions*, i.e., on pressure and concentrations of the residual-gas components, usually H_2, He, O_2, N_2, H_2O, CO, CO_2, CH_4, and Ar, in the accelerator, as well as on the ion energy and the charge-changing cross sections of beam ions colliding with the residual-gas (rest-gas) atoms

© Springer International Publishing AG 2018
I. Tolstikhina et al., *Basic Atomic Interactions of Accelerated Heavy Ions in Matter*, Springer Series on Atomic, Optical, and Plasma Physics 98,
https://doi.org/10.1007/978-3-319-74992-1_9

and molecules. For estimation of the vacuum conditions, the concentrations Y of the 'reference' atoms and molecules (H_2, N_2, and Ar) are often used with the following proportions: $Y(H_2) \approx 70\text{–}90\%$, $Y(N_2) \approx 20\text{–}30\%$, and $Y(Ar) \approx 1\text{–}3\%$.

The total ion-beam lifetime in accelerator consists mainly of three parts due to interactions with a residual gas, a gaseous target and cooling electrons (see [489, 490]). To estimate the ion-beam lifetime in cases of residual- and target-gas atoms and molecules, the following formula is commonly used:

$$\tau = \left(\rho \beta c \sum_T Y_T \left[\sigma_{EC}^{tot}(q, \upsilon, Z_T) + \sigma_{EL}^{tot}(q, \upsilon, Z_T) \right] \right)^{-1}, \quad \sum_T Y_T = 1, \quad (9.2)$$

where ρ denotes the gas density, q and υ the charge and velocity of the projectile, $\beta = \upsilon/c$, Z_T and Y_T the atomic number and fraction of the gas component T, σ_{EC}^{tot} and σ_{EL}^{tot} the *total*, i.e., accounting for the multiple-electron processes in electron-capture and electron-loss cross sections in collisions with target-gas particles. In the case of molecular targets, charge-changing cross sections are obtained by the Bragg's additivity rule. According to (9.2), at relativistic energies the ion-beam lifetimes turns to constant value because at $\upsilon \to c$ relativistic limit electron-loss cross sections also turns to constant values (see Sect. 6.4).

In real experimental conditions, the rest-gas density ρ and concentrations Y_T are different at different points of the accelerator volume and are also time-dependent. Moreover, the rest-gas atoms and molecules can be ionized by the beam ions, leading to a change in their interactions with the projectiles, thus, dynamic vacuum effects arise in the accelerator at very high beam densities [224]. All these circumstances restrict the application of (9.2) but, as a rule, it gives quite satisfactory results for estimating the ion-beam lifetimes in accelerators and storage rings.

As an example, experimental U^{28+}-ion beam lifetimes at specified vacuum conditions are shown in Fig. 9.1 as a function of ion energy in comparison with theoretical

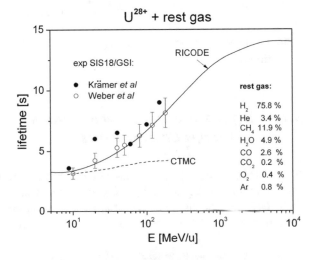

Fig. 9.1 U^{28+}-ion beam lifetimes as a function of ion energy at a gas pressure in the vacuum chamber about 10^{-10} mbar and rest-gas composition shown in the figure. Experiment: black circles—[491], white circles—[99]. Theory: (9.2) with electron-loss cross sections obtained with the CTMC code—dashed curve with RICODE program—solid curve. From [113]

calculations (see [113]). The vacuum parameters employed in the calculations are indicated in the figure. In calculations, the Bragg's additivity rule was used for loss and capture cross sections in collisions with molecules. The figure demonstrates quite good agreement between theory and experiment performed at the SIS18 synchrotron heavy-ion source at GSI, Darmstadt. We note that in the energy range considered, $E > 9$ MeV/u, the electron-capture cross sections of U^{28+} ions colliding with the rest-gas atoms and molecules are negligible compared to electron-loss cross sections. For relativistic energies, $E > 5$ GeV/u, the calculated lifetime of the uranium beam is predicted to be a constant value of about $\tau \approx 13$ s for the vacuum parameters considered.

9.1.2 Inverse Problem for Determination of the Vacuum Conditions in Accelerators

As is seen from (9.1), the ion-beam lifetime in accelerators depends on the vacuum parameters and charge-changing cross sections. In practice, however, simultaneous measurement of ion-beam lifetimes and residual-gas density and concentrations is a complicated problem, but one can solve the *inverse* problem to estimate the vacuum conditions from ion lifetimes and charge-changing cross sections known from theory or experiment.

Lifetimes τ of 11.4-MeV/u U^{q+}-ion beam are presented in Fig. 9.2 as a function of ion charge q. Experimental data on τ were obtained at the SIS18 heavy-ion synchrotron, Darmstadt, for charges $q = 34$–42, but the vacuum conditions were not known properly (see report [492]). Based on charge-changing cross sections calculated by the CAPTURE, DEPOSIT, and RICODE programs (see [43] for detailed description of the codes), and experimental data on τ, the vacuum para-

Fig. 9.2 11.4-MeV/u U^{q+}-ion beam lifetimes as a function of ion charge q. Experiment: filled and open circles—SIS18 data, Darmstadt [492]. Theory: dashed curve—calculation with electron-capture processes neglected, and solid curve—calculation with both electron-capture and electron-loss processes included. The vacuum parameters obtained with the help of calculations are indicated in the figure. From [42]

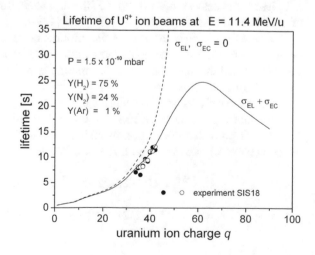

meters were estimated as follows: $\rho \approx 1.5 \times 10^{-10}$ mbar, $Y(H_2) \approx 75\%$, $Y(N_2) \approx 24\%$, $Y(Ar) \approx 1\%$, i.e., a good agreement with experimental ion-beam lifetimes was achieved with these vacuum conditions and calculated cross sections. The vacuum parameters obtained are close to expected parameters at the SIS18 synchrotron ion source.

Here we have to stress a key role of *electron-capture* processes at energy $E = 11.4$ MeV/u: inclusion of the capture processes leads to a decrease in the ion-beam lifetimes for uranium ions with charges $q > 60$, and neglecting the capture processes causes an infinite increase in ion-beam lifetime with increasing ion charge. Therefore, the uranium U^{q+} ions with charges $q \approx 60$ are the best candidates for detection of the longest ion-beam lifetime $\tau \approx 25$ s at $E = 11.4$ MeV/u.

Certainly, the solution of the inverse problem considered here is not unambiguous, but the procedure for estimating the vacuum parameters may be useful for interpretation of experimental data and planning future experiments with heavy many-electron ion beams in accelerator facilities.

9.2   Charge Exchange as a Mechanism for Creating an Inverse Population in a Capillary Discharge Plasma

A lasing on the Balmer-α line of H-like ions of carbon (18.22 nm) and oxygen (10.24 nm) in the soft X-ray band (in the transition of $n = 3 \rightarrow n = 2$) was identified in [493] and [494] using a low-inductance ablative discharge in a polyacetal capillary, where n is the principal quantum number. These results were obtained at the Ruhr University of Bochum, Germany, using the set-up described in [495].

Studies on inverted population in the soft x-ray spectral region have been carried out since 1985 in many laboratories including Livermore and Princeton, using powerful lasers and heavy targets ([496, 497]). At present, this problem is still a subject of high attention (see, for example, [498] and [499]).

Results of [493, 494] are of a special importance because the lasing in the x-ray range was obtained in a compact laboratory setup (table-top) using *light* ions. It should be noted that the first studies on creating inversion in the capillary discharge were performed in [500].

As it was concluded in [493, 494], a hot plasma of fully stripped C^{6+} and O^{8+} ions, produced in the necks of a $m = 0$ (hose) instability, streamed into the colder regions and produced population inversion by selective *ion-ion* charge exchange (electron capture) into the $n = 3$ levels of C^{5+} and O^{7+} ions:

$$C^{6+} + C^{2+}(2s2p) \rightarrow C^{5+}(n = 3) + C^{3+}, \tag{9.3}$$

$$O^{8+} + O^{3+}(2s2p\ ^2P) \rightarrow O^{7+}(n = 3) + O^{4+}. \tag{9.4}$$

A series of thin discs of population inversion along the axis was thus created leading to the amplification of the spontaneous emission on the Balmer-α line.

The interpretation of the observed lasing is supported by experiments with colliding laser-produced plasmas as well as theoretical calculations [494] of low-energy charge exchange cross sections and respective collisional-radiative calculations of excited level populations [501]. In [494] the charge exchange cross sections are calculated for collisions of oxygen bare nuclei with oxygen ions of lower ionization degrees that exist in cold regions of the plasma. The calculations are performed in the *adiabatic approach* using the ARSENY code [308] (see Chap. 5 in detail).

Figure 9.3 shows the n-resolved charge exchange cross section in collisions of bare oxygen with O^{4+} ions in the states $2s2p\ ^1P$ (Fig. 9.3a) and $2s2p\ ^3P$ (Fig. 9.3b) as a function of collision energy. As seen from the figure, the cross sections of all channels of these processes decrease rapidly at the energy $E \sim 0.1\,\text{keV/u}$.

The charge exchange cross sections of the process (9.4) reveal different behaviour: there is a reaction channel with final state $O^{7+}(n = 3)$ which has large cross sections

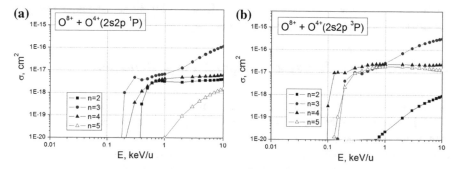

Fig. 9.3 Calculated ion-ion low-energy charge exchange cross sections for processes $O^{8+} + O^{4+}(2s2p\ ^{1,3}P) \to O^{7+}(n) + O^{5+}$, $n = 2\text{--}5$. From [494]

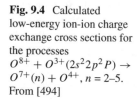

Fig. 9.4 Calculated low-energy ion-ion charge exchange cross sections for the processes $O^{8+} + O^{3+}(2s^2 2p^2 P) \to O^{7+}(n) + O^{4+}$, $n = 2\text{--}5$. From [494]

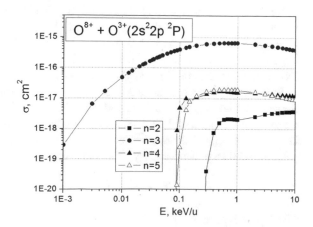

Fig. 9.5 Experimental time evolution of the current in the capillary discharge (Current I) and the Balmer-α line emission (PM) in the O^{7+} ion [494]. From [494]

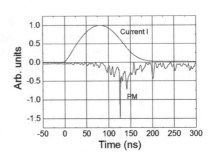

for low-energy collisions (Fig. 9.4). This reaction produces the population inversion causing the lasing at Balmer-α line.

Figure 9.5 shows a typical example of the time evolution of the current in the capillary discharge (current I) and the emission of the Balmer-α line of the O^{7+} ion at 10.24 nm detected with a photomultiplier (PM). At a time around 125 ns, a sharp deep peak is observed on the photomultiplier curve, which is interpreted as the moment of the population inversion [494]. The results of [494] show that lasing on the Balmer-α line of O^{7+} in a capillary discharge by charge exchange pumping during a $m = 0$ instability is indeed possible.

9.3 Detection of Heavy and Super-Heavy Elements

Uranium is the heaviest natural chemical element on Earth (atomic number $Z = 92$), and heavier elements, the so-called *super-heavy* elements (SHE), are produced artificially by nuclear fusion reaction of two elements. SHEs exist quite a short time and decay by different ways. In the previous years (2010–2016) the heaviest elements with $Z = 113$–118 were synthesized at the world powerful accelerators. A list of some heavy and super-heavy elements and their characteristics is given in Table 9.1. Properties of heavy and super-heavy elements are of great interests in atomic physics (structure of electronic shells, QED effects), quantum chemistry, and, naturally, in nuclear physics in studying nuclear shells, stability of isotopes, searching for the island of stability etc. (see [15, 503, 504]). Intensive investigations on creation and detection of SHE are carrying out in JINR (Dubna, Russia), RIKEN (Japan), Berkeley National Laboratory and Oak Ridge National Laboratory (USA) and in GSI (Darmstadt, Germany) (see [504]).

Super-heavy elements are produced in collisions of ion beams with foils of heavy elements at energies about a few hundreds of keV/u, where rates of nuclear reactions are maximal. For example, to synthesize isotope of Copernicium ion (Cn, $Z = 112$, $M = 277$ a.m.u.), ion-beam of ^{70}Zn ($Z = 30$, $M = 70$ a.m.u.) penetrates a lead foil ($Z = 82$, $M = 208$ a.m.u.) at energy ~ 350 keV/u:

$$^{70}Zn^{p+} +^{208} Pb \rightarrow^{277} Cn^{q+} + n, \tag{9.5}$$

where p and q are charge states.

Table 9.1 List of some heavy and super-heavy elements with atomic numbers $Z > 80$ in the periodic table

Z	Symbol	Name	Mass in a.m.u.	Electronic configuration	IP, eV	Half-life
80	Hg	Mercury	201	$4f^{14}5p^65d^{10}6s^2$	10.44	–
82	Pb	Lead	207	$5p^65d^{10}6s^26p^2$	7.42	–
83	Bi	Bismuth	209	$5p^65d^{10}6s^26p^3$	7.29	2×10^{19} y
87	Fr	Francium	223	$5d^{10}6s^26p^67s^1$	4.07	22 min
88	Ra	Radium	226	$5d^{10}6s^26p^67s^2$	5.28	1600 y
89	Ac	Actinium	227	$6s^26p^67s^26d^1$	5.38	21.77 y
92	U	Uranium	238	$6p^65f^36d^17s^2$	6.19	4.5×10^9 y
98	Cf	Californium	251	$6s^26p^65f^{10}7s^2$	6.28	900 y
100	Fm	Fermium	257	$6s^26p^65f^{12}7s^2$	6.50	100.5 d
102	No	Nobelium	259	$6s^26p^65f^{14}7s^2$	6.63	58 min
103	Lr	Lawrencium	266	$6p6\,5f^{14}7s^27p^1$	4.90	11 h
104	Rf	Rutherfordium	267	$6p^65f^{14}6d^27s^2$	6.01	1.3 h
109	Mt	Meitnerium	278	$6p^65f^{14}6d^77s^2$	9.55	7.6 s
110	Ds	Darmstadtium	281	$6p^65f^{14}6d^87s^2$	10.38	3.7 min
111	Rg	Roentgenium	282	$6p^65f^{14}6d^97s^2$	11.21	2 min
112	Cn	Copernicium	285	$6p^65f^{14}6d^{10}7s^2$	12.03	8.9 min
113	Nh	Nihonium	286	$5f^{14}6d^{10}7s^27p^1$	4.10	19.6 s
114	Fl	Flerovium	289	$5f^{14}6d^{10}7s^27p^2$	8.54	1.1 min
115	Mc	Moscovium	289	$5f^{14}6d^{10}7s^27p^3$	5.58	220 ms
116	Lv	Livermorium	293	$5f^{14}6d^{10}7s^27p^4$	6.69	61 ms
117	Ts	Tennessine	294	$5f^{14}6d^{10}7s^27p^5$	7.64	78 ms
118	Og	Oganesson	294	$5f^{14}6d^{10}7s^27p^6$	8.32	890 μs
119	–	(Uue)	–	$6d^{10}7s^27p^68s^1$	4.79	
120	–	(Ubn)	–	$6d^{10}7s^27p^68s^2$	5.85	

For each element, its symbol, name, mass in a.m.u. (M), electronic configuration of four outer-most shells(El. conf.), ionization potential (I) and reference of experiment (Ref. exp.) or theory (Ref. theory) where IP was obtained. Also half-life of the most stable isotopes from [502]. Elements Uue and Ubn have no names yet and which in Latin mean 119-th and 120-th. From [43]

One of the main principles used in detection of SHE is based on the properties of *equilibrium* charge-state fractions, which do not depend on the charge state of initial ion beam (see Sect. 3.3). This property of *atomic* interaction of ion beams with matter became the basis of the method to detect heavy and super-heavy elements. For detection of super heavy elements, *gas-filled separators* are used filled with H_2 or He, or their mixture at a pressure of about a few mbar ([15, 504–511]).

Experimentally, a distribution of the charge-state fraction F_q over q for the created ions (^{277}Cn) is unknown. However, if these ions are directed into gas-filled separator, i.e., a collision gas camera, then at a certain distance in the camera, the charge-state distribution becomes equilibrium with an equilibrium mean charge given by (3.6):

$$\bar{q} = \sum_q q F_q(\infty), \quad \sum_q F_q(\infty) = 1, \tag{9.6}$$

where $\bar{q}$ is independent of the initial charge-state distribution of the incoming (^{277}Cn) ions and the equilibrium distribution $F_q(\infty)$ is quite narrow consisted of only a small number of fractions.

If a mean charge of SHE in question is known, using the Lorentz formula it is easy to obtain the required *magnetic rigidity* of the separator dipole for detection of SHE with given $\bar{q}$ value:

$$B\rho = Mv/q, \tag{9.7}$$

where B, ρ, M and v denote a magnetic field, radius of circular path, mass and velocity of the ion. Therefore, knowing the mean charge of exit ions it is possible to find the magnetic rigidity of the dipole magnet required for detection of the given SHE, i.e., to find a required value of the magnetic field.

A knowledge about magnetic rigidity $B\rho$ is a key question in such experiments because a number of created super-heavy ions is very small and sharply decreases with the atomic number Z: a rate of creation of synthesized ions can vary from a few thousand ions per day for rather light ions to one ion per a few weeks (!) for super-heavy elements.

For a rough estimate of the mean charge for a given element with atomic number Z, the Bohr formula (3.19) is usually used:

$$\bar{q} = v Z^{1/3}, \quad 1 < v < Z^{2/3}, \tag{9.8}$$

which is valid for large Z and medium ion velocities v. As was mentioned before, the Bohr formula (9.8) and other semi-classical and semiempirical formulae have some disadvantages: they do not take into account the atomic structure of colliding particles and influence of the target-density effect, i.e., dependence on the gas pressure in the separator. Now it is known that these effects are very important and require introduction semiempirical corrections to (9.8) (see [15, 507–511]).

Figure 9.6 shows experimental data [507] of the mean charges for heavy and super-heavy elements $Z = 89$–116 obtained in a H_2-filled separator as a function of ion

Fig. 9.6 Experimental dependence of the equilibrium mean (average) charge for ions with atomic numbers $Z = 89$–116 as a function of ion velocity v at a pressure of 1 Torr of H_2-filled gas. v_0 is the Bohr velocity. The data were obtained at the DGFRS separator—Dubna Gas-Filled Recoil Separator. The straight line shows a linear fit of the experimental data to (9.9). The upper part of the systematics is shown in the inset where the average charges of No ($Z = 102$) at lower (0.5 Torr) and larger (1.5 Torr) pressures are shown by open circles. From [507]

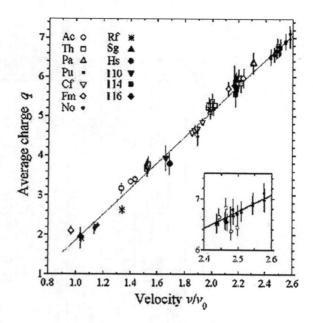

velocity $v = 1$–2.6 a.u. and a pressure $P = 1$ Torr. These data are well approximated by a formula similar to (9.8):

$$\bar{q} \approx 3.26v - 1.39. \qquad (9.9)$$

As seen from the inset of Fig. 9.6, experimental $\bar{q}$ data differ from linear dependence (9.9) because (9.8) does not take the density effect into account.

Similar measurements of the mean charges $\bar{q}$ for heavy and super-heavy ions with atomic numbers $Z = 80$–114 were carried out at TASCA separator (TransActinide Separator and Chemistry Apparatus) at GSI, Darmstadt, with He gas at pressures $P = 0.2$–2.0 mbar and energies of about a few hundred keV/u [510]. The result of the measured $\bar{q}$ values are given in Table 9.2 in comparison with the Bohr (9.8) and semiempirical formula, and with results of *atomic* calculations also presented in [510].

Values $\bar{q}_{SE1}$ were obtained by semiempirical formula obtained in [508] for SHE up to Rg ($Z = 111$) using H_2 gas in separator at a pressure of 0.66 mbar. Using the formula for $\bar{q}_{SE1}$ for $Z = 117$, the mean charge $\bar{q} \approx 6.8$ was predicted in [510] at He-gas pressure of 0.8 mbar. This predicted value was later used in [511] for detection of 117-th elements, called now *Tennessine* (see Table 9.1), at TASCA separator with magnetic rigidity $B\rho = 2.20$ Tm.

Values $\bar{q}_{SE2}$ in Table 9.2 were estimated by semiempirical formula from [202], obtained by analysis of experimental data for ions with $Z = 1$–92 and gaseous targets with atomic numbers $Z_T = 1$–54. As seen from the Table, the Bohr formula overesti-

Table 9.2 Experimental and calculated mean charge states $\bar{q}$ of heavy and super-heavy ions

ER	Z	Reaction	v (a.u.)	P(mbar)	$\bar{q}_{exp}$	$\bar{q}_B$	$\bar{q}_{SE1}$	$\bar{q}_{SE2}$	$\bar{q}_{th}$
^{180}Hg	80	^{144}Sm(^{40}Ar,4n)	2.84	0.6	6.97 ± 0.30	12.2	7.40	9.07	6.04
^{188}Pb	82	^{144}Sm(^{48}Ca,4n)	3.22	0.8	8.45 ± 0.19	14.0	8.60	10.55	7.83
^{205,206}Fr	87	^{181}Ta(^{30}Si,3-4n)	2.04	0.5	5.67 ± 0.19	9.0	6.06	6.46	5.96
$^{209-211}$Ra	88	^{158,160}Gd(^{54}Cr,3-4n)	3.17	0.6	9.37 ± 0.31	14.1	9.22	10.47	8.05
^{215}Ac	89	^{179}Au(^{22}Ne,4n)	1.39	0.8	4.28 ± 0.42	6.2	4.20	4.22	5.75
^{221,222}U	92	^{176}Yb(^{50}Ti,4-5n)	2.89	0.8	8.76 ± 0.29	13.0	8.64	9.80	8.27
^{252,254}No	102	^{206,208}Pb(^{48}Ca,2n)	2.40	0.8	6.68 ± 0.18	11.2	6.57	8.30	7.23
$^{254-255}$Rf	104	^{206,208}Pb(^{50}Ti,1-2n)	2.65	0.8	7.32 ± 0.25	12.5	7.30	9.37	7.02
^{288}Fl	114	^{244}Pu(^{48}Ca,4n)	2.30	0.8	6.70 ± 0.37	11.1	6.97	8.28	8.02
^{287,288}Uup	115	^{243}Am(^{48}Ca,3-4n)	2.28	0.8		11.1	7.03	8.23	7.70
^{293,294}Uus	117	^{249}Bk(^{48}Ca,3-4n)	2.25	0.8		11.0	7.16	8.19	8.58
^{293,294}Uue	119	^{249}Bk(^{50}Ti,3-4n)	2.42	0.8		11.9	7.83	8.92	8.73
^{293,294}Ubn	120	^{249}Cf(^{50}Ti,3-4n)	2.43	0.8		12.0	7.93	9.02	9.03

The identified evaporation residues (ER), i.e., reaction products, atomic numbers (Z), nuclear reactions, ion velocities and the measured He gas pressures are also given. The results of Bohr's predictions (9.8) and two different semiempirical approaches $\bar{q}_{SE1}$ and $\bar{q}_{SE2}$ [202] are presented together with theoretical $\bar{q}_{th}$ results. From [510]

mates results by a factor of 2; a little better agreement is achieved using a formula for $\bar{q}_{SE1}$ and the best agreement with experimental data is obtained with $\bar{q}_{SE2}$ formula.

It is well known that determination of magnetic rigidity by known mean charge is a quite complicated experimental problem, which is usually solved semiempirically by calibration of a gas-filled separator on the mean charges of isotopes of *stable* elements with account for specific properties of specific experimental device.

As has been mentioned before, calculation of the mean charges $\bar{q}$ for heavy and super-heavy elements with $Z = 80$–120 in [510] was for the first time performed on the basis of *atomic* calculations of electron-loss and electron-capture cross sections with the density effect accounted for using the balance rate equations (3.1) and (3.2) for charge-state evolution. The data calculated this way are marked $\bar{q}_{th}$ in Table 9.2. As is seen the $\bar{q}_{th}$ values agree with experimental data within 20%. However, for precise measurements of super-heavy elements this accuracy of $\bar{q}$ values is not sufficient because the required accuracy should be of about a few percent (see [510]).

9.4 Effects of Charge-State Evolutions on Material Modifications Using Swift Heavy Ions

As has been discussed in the previous chapters, when even a single fast (swift) heavy ion in MeV/u energy range impinges a target medium, unique characteristic features, which cannot be brought about by any other means, like photon or electron impacts, are exhibited as a result of cumulative effects of a number of consecutive elastic and inelastic collisions between the projectile ion and target atoms. Such interactions of energetic ions with materials constitute the basis of a wide range of applications, like materials analysis, materials modification and so on, and are considered as an ideal tool to alter and control material properties through atomic structure modification of the target material.

In collisions of swift heavy ions, energies transferred to target electrons via inelastic processes are much larger than those to target nucleus via elastic collisions by three orders of magnitude. Such target-electron excitations are considered to play a role in materials modification, as they take place as deep as an ion range of 10 μm inside the medium. Since each inelastic collision process is significantly affected by charge and electronic states of the projectile ions, information on the distribution and evolution of charge states in matter is essential to basic study and applications of heavy-ion irradiations. The energies of the excited target-electrons are finally transferred to the target lattice and provide ultrafast local heating along the ion path, through which a cylindrical damage region of several nm of diameter (ion track) is formed when the electronic energy is larger than a material-dependent threshold value [512].

A number of experimental efforts have been devoted to material modifications via swift-heavy-ion irradiations, in most of the cases however, charge states of the impinging ions were not so much cared as about ion energies, i.e., the charge states were selected according to maximum terminal voltage of accelerators and desired

Fig. 9.7 X-ray diffraction (XRD) intensity degradation of WO_3 films irradiated by 90-MeV Ni^{10+} and 100-MeV Xe^{14+} ions normalized to those irradiated by those ions with equilibrium charge states. From [517]

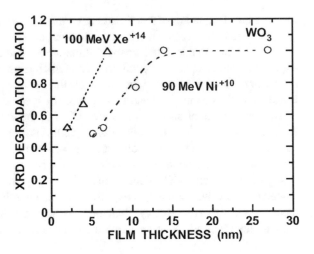

ion energies. In several recent experiments, it has been proved that ion charge state plays an important role in material modifications in the near-surface area, such as electron or photon emission from surfaces or formation of craters and ion tracks in polymers [513–516].

Lately, Matsunami et al. [517] have studied atomic structure modification of thin WO_3 films irradiated by 90-MeV Ni^{10+} and 100-MeV Xe^{14+} ions with and without a C-film of 4.7×10^{12} atoms/cm^2 upstream the target, i.e., by Ni/Xe ion beams with charge state 10+/14+ and with their equilibrium charge states which is higher than 10+/14+. An X-ray diffraction (XRD) observation of the irradiated films has shown that XRD intensities decrease with ion irradiation and the XRD intensity decrease is more pronounced for the irradiation with equilibrium-charge-state beams than with 10+/14+ beams. It has also been observed that the XRD intensity degradations for Ni^{10+} and Xe^{14+} irradiations depend on WO_3 film thickness, whereas those for irradiation with equilibrium-charge-state beams are nearly constant for different WO_3 target thickness. Those XRD intensity degradations for Ni^{10+} and Xe^{14+} irradiations normalized to those for irradiations with equilibrium-charge-state beams are plotted in Fig. 9.7. It is obvious that the normalized XRD intensity degradations increase and finally saturate to the unity as the WO_3 film thickness grows, clearly demonstrating that the material property modification is affected by the evolution and equilibration of impinging ion charge-states inside the target. Detailed discussion of charge-state and stopping-power evolutions is made in [517].

Chapter 10
Lifetimes of Radioactive Heavy Ion Beams

Abstract In the previous chapter, it was discussed that for *non-radioactive* ion beams, the lifetimes depend on the atomic charge-changing interactions between the ion beam and residual-gas components. In the case of radioactive ion beams, the mean lifetimes depend in addition on the *nuclear* properties of accelerated ions, and hence, the total mean lifetime comprises both atomic and nuclear components:

$$\frac{1}{\tau} = \frac{1}{\tau_{atomic}} + \frac{1}{\tau_{nuclear}}, \tag{10.1}$$

or equivalently:

$$\lambda = \lambda_{atomic} + \lambda_{nuclear}, \tag{10.2}$$

where $\lambda = 1/\tau$ is the total decay rate. Each decay components may have several terms, which correspond to different decay channels (e.g., $\alpha-$, β^+-, β^--decay, fission, etc.) and/or state transitions (e.g., nuclear ground-ground state transitions, ground-excited state transitions, etc.), so one can define:

$$\lambda_{nuclear} = \sum_{\text{all channels } i} \lambda_i. \tag{10.3}$$

The individual decay rate depends on the underlying processes. In the following sections, we describe mechanisms of the decay rates, or mean lifetimes, due to the nuclear reactions.

10.1 The α-Decay Mean Lifetime of Heavy Ions

The α-decay half-lives of heavy radioactive ions are varying from a few microseconds to about 10^{17} s. Alpha emitters can be found for elements heavier than Bi ($Z = 83$), or for rare-earth elements from Nd ($Z = 60$) to Lu ($Z = 71$). Some examples of alpha emitters are shown in Table 10.1.

The decay rate (mean lifetime) can be estimated using a model of a α particle, bouncing inside a box (potential well) at a frequency υ/R and having a probability

© Springer International Publishing AG 2018
I. Tolstikhina et al., *Basic Atomic Interactions of Accelerated Heavy Ions in Matter*, Springer Series on Atomic, Optical, and Plasma Physics 98, https://doi.org/10.1007/978-3-319-74992-1_10

Table 10.1 Half-lives of heavy and super-heavy atoms due to the α-decay

Reaction	Half-life
$^{238}_{92}\text{U} \rightarrow\ ^{234}_{90}\text{Th} + \alpha$	4.5 gigayears
$^{218}_{84}\text{Po} \rightarrow\ ^{214}_{82}\text{Pb} + \alpha$	3.05 min
$^{214}_{84}\text{Po} \rightarrow\ ^{210}_{82}\text{Pb} + \alpha$	16.37 ms
$^{222}_{86}\text{Rn} \rightarrow\ ^{218}_{84}\text{Po} + \alpha$	3.82 days

Fig. 10.1 Quantum tunneling of a α-particle inside a potential well. From [528]

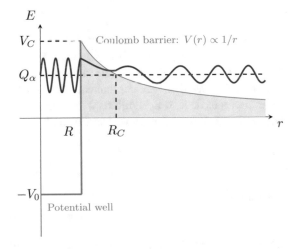

of quantum tunneling (Fig. 10.1) via potential barrier in the form:

$$\lambda_\alpha = \frac{\upsilon}{R} e^{-2G}, \tag{10.4}$$

where υ denotes the α-particle velocity inside a nucleus, R the potential well size, and $\exp(-2G)$ the probability of tunneling through the potential well. The factor G is the so-called Gamow factor, which is defined as:

$$G = \frac{1}{2}\sqrt{\frac{E_G}{Q_\alpha}}\, g\left(\sqrt{\frac{R}{R_c}}\right), \tag{10.5}$$

where $g(x) = \frac{2}{\pi}\left(\arccos(x) - x\sqrt{1 - x^2}\right)$, R_c is the distance from origin to the point, at which the Coulomb potential V_C is equal to the value Q_α for the α decay reaction:

$$R_c = \frac{e^2 Z_\alpha Z}{Q_\alpha}, \tag{10.6}$$

Fig. 10.2 Plot of the
Geiger-Nuttall law ($\log(\lambda)$
versus $Z/\sqrt{Q_\alpha}$) with
$a_1 \approx 128$ and $a_2 \approx 3.97$

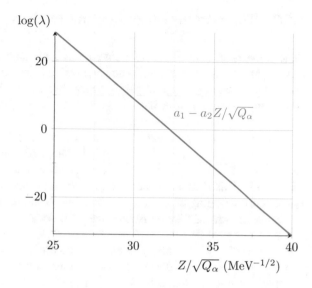

Here $Q_\alpha = m_{\text{parent}} - m_{\text{Daughter}} - m_\alpha$ and E_G is the Gamow energy [518]:

$$E_G = \left(\frac{2\pi Z_\alpha Z e^2}{\hbar c}\right)^2 \frac{\mu c^2}{2} \tag{10.7}$$

where Z and Z_α denote the atomic numbers of the two decay products, the α particle and the daughter nucleus, respectively, and μ the reduced mass. We note that this results is also valid for spontaneous fission if we substitute Z_α, and Q_α to the above equations with the corresponding quantities of the fission products. For example, considering the ^{238}U $\rightarrow$ ^{234}Th $+ \alpha$ decay, we find $E_G \approx 122\,\text{GeV}$, $\sqrt{E_G/Q_\alpha} \approx 171$, $g(\sqrt{(R/R_C)}) \approx 0.518$, which gives $e^{-2G} \approx 4 \times 10^{-39}$. For the bounce frequency $\upsilon/R \approx 4.3 \times 10^{21}\text{s}^{-1}$, one has for the half-life $T_{1/2} = 4.5\,\text{Gy}$ (gigayears).

The above results for α decay can be approximate and further reduced into a simpler form (see Fig. 10.2), known as the Geiger-Nuttall law [519]:

$$\ln(\lambda_\alpha) = a - b\frac{Z}{\sqrt{Q_\alpha}}, \tag{10.8}$$

where a and b are nearly constants. For example, in the approximation of $\mu \approx m_\alpha$ and $R \ll R_c$ and thus $g(\sqrt{R/R_c}) \approx 1$, we obtain $b = 2\pi\alpha\sqrt{2m_\alpha c^2} = 3.97\,\text{MeV}^{1/2}$, where $\alpha = e^2/(\hbar c)$ is the fine structure constant. The parameter a varies around the value of 128.

10.2   The β-Decay Mean Lifetime of Heavy Ions

The β-decay stem from the weak interaction and can be described—at the nuclear
level—as the transformation of a nucleon inside a nucleus, e.g., neutron into proton,
with the emission of a lepton pair (electron-neutrino). This is an isobaric transition
(constant mass number A but change in the (Z, N) pair). The β decay occurs as
soon as the Q-value of the reaction is positive. This can be better seen in an isobar
cut across the valley of stability as shown in Fig. 10.3. For isobars, the binding
energy as a function of Z follow a parabolic shape (BetheWeizsäcker semi-empirical
mass formula). The Q-value is roughly equal to the parent-child binding energy
difference. This difference increase linearly, and then we move from the stability
point, corresponding to increasing β instability (shorter mean lifetime).

There are several distinguished forms of β decay:

$n \rightarrow p + e^- + \overline{\nu_e}$ β^- decay, emits e^- to continuum
$p \rightarrow n + e^+ + \nu_e$ β^+ decay, emits e^+ to continuum
$p + e_b^- \rightarrow n + \nu_e$ EC, orbital electron capture
$n \rightarrow p + e_b^- + \overline{\nu_e}$ bound state $\beta -$ decay (β_b), emits e^- to a free electron orbit

We note that the orbital electron capture and the bound-state β-decay have a
two-body decay kinematics, in which the (anti) neutrinos are emitted with a well-
defined energy. Note also, that electron capture from continuum can only occur at
high electron densities like, for example, in a dense stellar plasma. The β decay
half lives range from few milliseconds to gigayears. Some examples are given in
Table 10.2.

Fig. 10.3 Illustration of an
isobar cut across the valley
of stability. The β-stable
isobars minimize the binding
energy, so that the transitions
$Z \rightarrow Z \pm 1$ have negative Q
values. The further an isobar
is from the stability point,
the higher is the decay rate

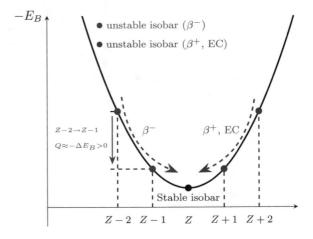

Table 10.2 Examples of some beta emitters, their transition type and half-lives

Reaction	Transition	Half-life
$_6^{14}C \rightarrow {_7^{14}}N + e^- + \bar{\nu}_e$	β^-	5730 years
$_2^6He \rightarrow {_3^6}Li + e^- + \bar{\nu}_e$	β^-	0.8 s
$_8^{14}O \rightarrow {_7^{14}}N + e^+ + \nu_e$	β^+	70.6 s
$_{26}^{55}Fe + e_b^- \rightarrow {_{25}^{55}}Mn + \nu_e$	EC	2.6 years
$_{27}^{57}Co + e_b^- \rightarrow {_{26}^{57}}Fe + \nu_e$	EC	271.8 days

10.2.1 Fermi Interaction

The decay-rate theory of β decay is usually described by a time-dependent perturbation theory, known as the universal Fermi interaction rule (Fermi's Golden rule
[520]). The probability of a decay (transition rate per unit time $\lambda_{i \rightarrow f}$) from the initial
energy eigenstate of a quantum system $|i\rangle$ into a final state $|f\rangle$ is given by a Fermi's
golden rule:

$$\lambda_{i \rightarrow f} = \frac{2\pi}{\hbar} \underbrace{|\langle i|H|f\rangle|^2}_{=|M_{if}|^2} \rho(Q), \qquad (10.9)$$

where H is the transition hamiltonian, $\rho(Q)$ is the density of final states, and Q
is the Q-value, the released energy of the reaction. The quantity $\langle i|H|f\rangle$ is the
transition matrix element M_{if}. Although the density of final state can be computed,
the transition matrix is usually difficult to compute, as it depends on many factors.
A short introduction to the β-decay theory is given in Appendix B. The description
of β decay, given in the Appendix, is the so-called *four fermions interaction*, which
is valid for low and medium energies (Fig. 10.4).

Fig. 10.4 Diagram of the
β^+ decay, a proton which
decays into a neutron with
the emission of a positron
and a neutrino. Time is going
from bottom to top

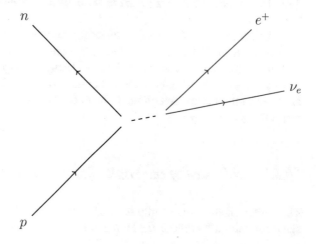

10.3   Beta Decay in Stellar Environment

In stellar interior, the distribution of the atomic states obeys the Boltzmann distribution (in case of non-degenerate matter), which is sensitive to the temperature. In extreme conditions like those, encountered in stars, thermal collisions of atoms lead to their ionization. The ejected electrons from atom form an electron gas, which co-exists with the gas of atomic ions and neutral atoms. The thermal-ionization rate per volume increases with the density number $n_{i,j}$, where i and j denote the atom species and its j-times ionized state, respectively. However, electron recombination occurs as well due to the interaction with the surrounding free electrons. The recombination rate per volume involves two particles—electron and the ionized atom—and increases with the product of their densities $n_{i,j+1} \cdot n_e$. At the local thermodynamical equilibrium, the population of differently ionized states $n_{i,j+1}/n_{i,j}$ is given by the Saha ionization equation:

$$
\frac{n_{i,j+1}\, n_e}{n_{i,j}} = 2 \frac{Z_{i,j+1}}{Z_{i,j}} \left(\frac{2\pi m_e}{h^2} \right)^{3/2} (k_B T)^{3/2} e^{-\chi_{i,j}/k_B T}, \tag{10.10}
$$

where $Z_{i,j}$ denote the atomic partition function, m_e the electron mass, h the Planck constant, k_B the Boltzmann constant and $\chi_{i,j}$ the ionization potential. We note that the charge neutrality has to satisfy the condition $n_e = \sum j n_{i,j}$.

At high density and pressure, free electrons and the neighboring ions can influence the potential distribution in and near a given ion. This effect leads to reduce the ionization potential $\chi_{i,j}$, compared to the one $\chi_{i,j}^{lab}$ measured in a laboratory:

$$
\chi_{i,j} = \chi_{i,j}^{lab} - \Delta_{ij}, \tag{10.11}
$$

where Δ_{ij} can be estimated theoretically [521]. A smaller ionization potential results in a shift of the equilibrium occupation number in the direction of increased ionization. This effect is called a *pressure ionization* or a *depressure of the continuum* [521].

Calculation of the K- and L-shell occupation numbers for Ho atoms have been performed in [521] for various temperature and mass density, taking into account both the thermal and pressure ionization. Using the s-process parameters, a mass density of $\rho_s \approx 8 \times 10^3$ g/cm^3 and temperature of $T_s \approx 3.5$ gives for K- and L-shell in Ho atom the occupation numbers from 0.22 up to 1.25 and from 0.1 up to 0.67, respectively.

10.3.1   Influence of the High Charge State on β-Decay

The high ionization state (fully ionized or H-like Ho ions in the previous example) can drastically affect the β-decay rates. It turned out that the decay modes, known in neutral atoms, can become forbidden, while the new ones can be opened up. For example, the energy release, $Q_{\beta_b}^{K,L}$, for β_b-decay to K, L, ...shells strongly depends on the electronic configuration. In the case of a fully ionized mother atom of nuclear charge number Z and atomic mass number A, which decay by β_b-decay to the K shell, one has:

$$m_{(Z,A)} \rightarrow m_{(Z+1,A)} + m_e - B_{Z+1}^{K,L}, \tag{10.12}$$

where $B_{Z+1}^{K,L}$ is the binding energy of the K- or L-electron in the H-like daughter with nuclear charge number $Z + 1$. In the case of β^--decay in neutral atom we have:

$$m_{(Z,A)} + Zm_e - B_Z^e \rightarrow m_{(Z+1,A)} + (Z + 1)m_e - B_{Z+1}^e \tag{10.13}$$

with the corresponding Q-value,

$$Q_{\beta^-} = m_{Z,A} - m_{Z+1,A} - m_e - (B_Z^e - B_{Z+1}^e), \tag{10.14}$$

where B_Z^e denotes the total binding energy of all bound electrons.

Defining $\Delta B = B_{Z+1}^e - B_Z^e$, we can write the relation between the Q-value for β_b-decay in fully ionized to the Q value for β^--decay in neutral atom as:

$$Q_{\beta_b}^{K,L} = Q_{\beta^-} - \Delta B + B_{Z+1}^{K,L}, \tag{10.15}$$

While values of ΔB range in neutral atoms from 64 eV to about 20 keV for low and high nuclear charge number Z, respectively, the K binding energy B_{Z+1}^K can reach values as large as 130 keV ($Z = 92$). From these considerations, it follows that β^--stable neutral atoms with negative Q_{β_c}-value can become β_b-unstable if

$$B_{Z+1}^{K,L} > |Q_{\beta^-}| + \Delta B, \tag{10.16}$$

which is realized for sufficiently small negative Q_{β^-} values. We note that for neutral parent atoms we have $Q_{\beta^-}^{(Z,A)} = -Q_{EC}^{(Z+1,A)}$, which means that a stable neutral atoms with a small Q_{β^-} value can be found in daughter atoms of EC-unstable parents with a small positive Q_{EC} value. It is clear that stable nuclei, which becomes unstable in stellar medium, may strongly influence a nucleosynthesis paths, and, in particular, the s-process path, which follows the stability valley. In general, such a dependence of decay properties on the atomic charge state can have substantial impact on the nucleosynthesis in hot stellar plasma [521, 522].

10.3.2   Influence of the Induced Transitions on the Stellar
β-Decay Rates

Excited nuclear states, in addition to ionization effect, are significantly populated by induced transitions due to the intense thermal photon bath. If one of the excited states decays faster than the ground state, the effective stellar β-decay rate can be much larger than the terrestrial rate as well [523]. The stellar β-decay rates become a combination, levelheaded via the probability p_i, of decay rates of all considered states towards the available states of the product nuclei:

$$\lambda_\beta^{star} = \sum_i \left(p_i \sum_j \lambda_{\beta_{ij}} \right), \tag{10.17}$$

where the probability p_i is the product of the statistical weight and the Boltzmann factor:

$$p_i = \frac{(2J_i + 1) \exp(-E_i/kT)}{\sum_m (2J_m + 1) \exp(-E_m/kT)} \tag{10.18}$$

The effects of ionization, temperature and density on the β-decay rates were carefully investigated in the comprehensive work of Takahashi and Yokoi [524].

10.4   Example: Orbital Electron Capture Decay of Stored Highly Charged Ions

As seen in Sect. 10.3.1, the β-decay rate can depend on the ionization state. This effect has been observed in some experiments [525, 526], performed at GSI, Darmstadt, which showed that the orbital electron-capture decay rates of H- and He-like ions can be larger than of neutral atoms having about 60 electrons. This effect has been observed in H- and He-like ^{140}Pr and ^{142}Pm ions using the Schottky Mass Spectrometry method. These nuclei have been chosen because of their nuclear properties. Indeed, these two nuclei decay by a pure Gamow-Teller transitions $1^+ \rightarrow 0^+$ to almost 100% (ground state to ground state), and thus, are ideal systems for experimental study of such transitions, as they have well defined initial and final states (see Fig. 10.5).

In both experiments, measurements were performed using the Fragment Separator (FRS) and the Experiment Storage Ring (ESR) facilities. The mean lifetime can be measured using time resolved Schottky Mass Spectrometry method. An illustration of a slice of the time-resolved spectrum with the different atomic and nuclear processes occurring during the measurements is shown in Fig. 10.6. Table 10.3 shows the results of the ^{142}Pm measurements of the β^+- and EC-decay rates, as well as the EC branching ratios for various charge states. A significant increase of the EC branching from about 23% in neutral atoms to 29% in the one-electron system is observed (Fig. 10.6).

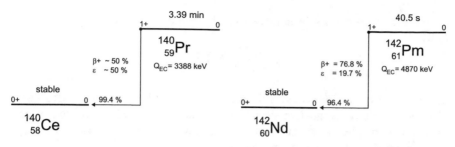

Fig. 10.5 Decay scheme of the ^{140}Pr and ^{142}Pm nuclei. Decay schemes of neutral ^{140}Pr and ^{142}Pm. Both nuclei decay by pure Gamow-Teller transition from ground state to ground state to almost 100%. In these transitions, neither an X-ray, nor γ-rays is emitted. Only a monochromatic electron neutrino and a corresponding monochromatic recoil of the daughter are left after the decay. From [528]

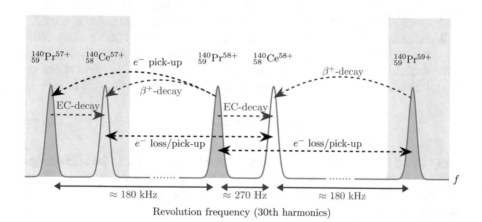

Fig. 10.6 Illustration of the various processes which can occur during a SMS measurement at ESR. In the experiments discussed above, mechanical slits placed on the appropriate orbits (shaded area) prevents any feeding coming from the β^+-daughter ions and/or any charge exchange reaction, so that only the ions of interest circulate in the storage ring. From [528]

Table 10.3 Averaged results compared to the measured rates of neutral atoms taken from [527]

Ion	$\lambda_{\beta^+}[s^{-1}]$	$\lambda_{EC}[s^{-1}]$	$EC/(\beta + EC)$
^{142}Pm^{61+} (bare)	0.01228 (70)	–	–
^{142}Pm^{60+} (H-like)	0.01257 (32)	0.00514 (14)	29.0% (13)
^{142}Pm^{59+} (He-like)	0.01393 (59)	0.00357 (10)	20.2% (10)
^{142}Pm (Neutral) [527]	0.01319 (49)	0.00392 (46)	22.9% (27)

The EC-decay rate ratio in both cases, i.e., ^{140}Pr and ^{142}Pm ions, has been estimated to as

$$\frac{\lambda_{EC}^{\text{H-like}}}{\lambda_{EC}^{\text{He-like}}} \approx 1.5, \qquad (10.19)$$

meaning that H-like ions decay 1.5 times faster than He-like ions. A behavior of electron density at the nucleus for these systems does not allow to explain these results. In principle, the electron density at the nucleus for an electron of shell n, is given in the plane wave approximation, by

$$|\psi_e^n(0)|^2 \approx \frac{1}{\pi} \left(\frac{m_e e^2}{4\pi\varepsilon_0 \hbar^2} \right)^3 \cdot \left(\frac{Z}{n} \right)^3, \qquad (10.20)$$

which is proportional to $1/n^3$ for a given nucleus. Therefore, these ratio should be of about 0.5 since the one- or two bound electrons of the ions are in the K-shell.

It has been shown [529] that this difference can be explained by taking into account the conservation of the total angular momentum of the system (nucleus+lepton) before and after the EC decay process. The general treatment of EC-decay in Gamow-Teller transition for H and He-like ions is given in [529] and leads to the ratio:

$$P_{EC}^H / P_{EC}^{He} = \frac{2I + 1}{2F + 1}, \qquad (10.21)$$

where F and I are the total angular momentum and nuclear spin of the initial states, respectively. With a nuclear spin $I = 1$ and a total angular momentum $F = 1/2$ one also finds a ratio of $3/2$, which is in agreement within one standard deviation with the measured ratio of 1.44 (6).

Appendix A
Solution of the Balance Rate Equations

In this Appendix, analytical solution of the balance rate equations (3.1) and (3.2) for the charge-state fractions $F_q(x)$ is found using the eigenvalue-decomposition method (see, e.g., [530]) provided all charge-changing cross sections are given.

A.1 Reducing the Dimension of the Balance Equation System

Let us first reduce the dimension of the equation system by one. The system of equations (3.1) and (3.2) for N charge states $q_1 \leq q \leq q_N$ can be written in the form:

$$\frac{d}{dx} F_{q_i}(x) = \sum_{q_j=q_1}^{q_{i-1}} \sigma_{q_j q_i}^{EL} F_{q_j}(x) + \sum_{q_s=q_{i+1}}^{q_N} \sigma_{q_s q_i}^{EC} F_{q_s}(x)$$

$$- \left(\sum_{q_m=q_{i+1}}^{q_N} \sigma_{q_i q_m}^{EL} + \sum_{q_k=q_1}^{q_{i-1}} \sigma_{q_i q_k}^{EC} \right) F_{q_i}(x), \tag{A.1}$$

$$q_1 \leq q_i \leq q_{N-1}, \tag{A.2}$$

$$\sum_{q=q_1}^{q_N} F_q = 1, \tag{A.3}$$

where EL and EC denote electron-loss and electron-capture cross sections, respectively.

The fraction F_{q_N} can be obtained from the normalization condition (A.3):

$$F_{q_N} = 1 - \sum_{q=q_1}^{q_{N-1}} F_q. \tag{A.4}$$

© Springer International Publishing AG 2018
I. Tolstikhina et al., *Basic Atomic Interactions of Accelerated Heavy Ions in Matter*, Springer Series on Atomic, Optical, and Plasma Physics 98, https://doi.org/10.1007/978-3-319-74992-1

Substituting (A.4) into (A.1), the equation system can be re-written as an $N - 1$ dimension system:

$$\frac{d}{dx} F_{q_k} = \sum_{q_i=q_1}^{q_{N-1}} a_{q_k,q_i} \, F_{q_i} + g_{q_k}, \tag{A.5}$$

where $q_1 \leq q_k \leq q_{N-1}$; a_{q_k,q_i} is an algebraic sum of the cross sections $\sigma_{q_i q_j}$ and can be both positive and negative, and $g_{q_k} = \sigma_{q_N q_k}^{EC}$ is the contribution of electron capture cross sections from the last state q_N into the state q_k.

A.2 Solving the System

One can easily recognize that (A.5) is a typical differential equation system with a constant term g_{q_k}. According to the Cauchy-Lipschitz theorem, for a given initial condition $F(x = 0) = F_0$ there exists a unique solution of the equation system (A.5), which can be written in the form:

$$F(x) = F^h(x) + Z(x), \tag{A.6}$$

where $Z(x)$ is the particular solution of (A.5) and $F^h(x)$ is the solution of the homogeneous equation

$$\frac{d}{dx} F^h = A F^h. \tag{A.7}$$

Here A denotes an $N - 1$ dimensional matrix and F the charge-state fraction vector.

A.2.1 Particular Solution

It is convenient to use as a particular solution the equilibrium fractions $Z = F(\infty)$, i.e., the solution of equation

$$A F(\infty) + g = 0. \tag{A.8}$$

The solution of (A.8) is trivial : $F(\infty) = -A^{-1} g$ and corresponds to the asymptotic solution of equations (A.1) and (A.3) at $x \to \infty$.

A.2.2 Homogeneous Solution

The homogeneous equations (A.7) can be solved using the eigenvalue decomposition method. If A can be diagonalized then one has

$$A = PDP^{-1}, \tag{A.9}$$

where D is a diagonal matrix with eigenvalues λ_i, and i being the index of the charge q_i, and $P = \{v_1, \ldots, v_{N-1}\}$ is the eigenvector matrix. After change of variable

$$Y = P^{-1}F^h \tag{A.10}$$

the equation (A.7) becomes:

$$\frac{d}{dx}Y = DY. \tag{A.11}$$

The solution to this equation has the form:

$$Y(x) = \begin{pmatrix} C_1 e^{\lambda_1 x} \\ \vdots \\ C_{N-1} e^{\lambda_{N-1} x} \end{pmatrix}, \tag{A.12}$$

where coefficients C_i are determined by the initial conditions:

$$Y(x = 0) = \begin{pmatrix} C_1 \\ \vdots \\ C_{N-1} \end{pmatrix} = P^{-1} F_0^h, \tag{A.13}$$

with $F_0^h = F_0 - F(\infty)$. The homogeneous solution is then obtained by

$$F^h(x) = PY(x). \tag{A.14}$$

Therefore, the solution of the rate equations is expressed as a linear combination of exponential functions with eigenvalues as exponents, and with coefficients that are proportional to the eigenvectors components. The eigenvalues and eigenvectors can eventually be complex.

A.2.3 Case of Complex Eigenvalues

If some of the obtained eigenvalues are complex, then they necessarily occur by pair of complex conjugates with their corresponding complex conjugate eigenvectors. In that case, it is always possible to find a new and real eigenvector basis by using the complex conjugates properties. To illustrate this case, we consider a simple example. Let us assume that we have only one pair of complex eigenvalues, with indexes 1 and 2, e.g., $\lambda_1 = \theta + \omega i$ and $\lambda_2 = \theta - \omega i$. Then the matrix $P = \{v_1, v_2, \ldots, v_{N-1}\}$ will have one pair of complex conjugate eigenvectors v_1 and v_2 and the solution will be

$$F^h = C_1 \Re(e^{i\omega x} v_1) e^{\theta x} + C_2 \Im(e^{i\omega x} v_1) e^{\theta x} + \sum_{i \neq 1,2} C_i e^{\lambda_i x} v_i, \qquad (A.15)$$

where $\Re$ and $\Im$ denote the real and imaginary parts, respectively.

Let us write the complex elements of the eigenvector v_1 as $a_k + ib_k$. Then, the complex eigenvectors component can be written as follows:

$$\Re(e^{i\omega x} v_1) e^{\theta x} = e^{\theta x} \begin{pmatrix} a_1 \cos \omega x - b_1 \sin \omega x \\ a_2 \cos \omega x - b_2 \sin \omega x \\ \vdots \\ a_{N-1} \cos \omega x - b_{N-1} \sin \omega x \end{pmatrix} \qquad (A.16)$$

$$\Im(e^{i\omega x} v_1) e^{\theta x} = e^{\theta x} \begin{pmatrix} a_1 \sin \omega x + b_1 \cos \omega x \\ a_2 \sin \omega x + b_2 \cos \omega x \\ \vdots \\ a_{N-1} \sin \omega x + b_{N-1} \cos \omega x \end{pmatrix}, \qquad (A.17)$$

and the unknown coefficients C_i can finally be determined by the initial condition:

$$Y(x = 0) = P_{\Re}^{-1} F_0^h,$$

where $P_{\Re} = \{\Re(v_1), \Im(v_1), v_3, \ldots, v_{N-1}\}$ is the new eigenvector matrix.

Appendix B
Introduction to the Theory of β-Decay Rate

As has been shown in Sect. 10.2, the β-decay rate is given by the universal Fermi theory, also called *four fermions interaction theory*. The decay rate expression is the product of two factors, namely the square matrix element $|M_{if}|^2$ and the density of final states $\rho(Q)$. The density of final states can be written as:

$$\rho(Q) = \frac{dN}{dQ} = \frac{dN_e dN_v}{dE_e dE_v},$$

where $dN = dN_e dN_v = (4\pi p_e^2 dp_e)(4\pi p_v^2 dp_v)$ is the product of the two elementary volumes in the phase space. By differentiation and by neglecting the recoil energy one can derive the following expression:

$$\rho(Q) = (4\pi)^2 E_e(Q - E_e)\sqrt{E_e^2 - m_e^2}\sqrt{(Q - E_e)^2 - m_v^2} dE_e \qquad \text{(B.1)}$$

where $Q = E_e + E_v$.

In the Fermi theory, one assumes that a vector current is coupled with a potential at the same point of the space-time via a "contact" interaction. Since in β-decay, the relativistic neutrinos and electrons have small masses compared to their kinetic energies, their wave functions have to be solutions of the Dirac equation. The vector current is defined as the hadronic vector current density, e.g. $J_-^\mu = (\overline{\psi_p}\gamma^\mu \psi_n)$ and the potential vector is constructed from the lepton field yielding the leptonic current density, e.g. $A_\mu = (\overline{\psi_e}\gamma_\mu \psi_v)$ where γ^μ denotes the Dirac matrices.[1]

---

[1] The Dirac matrices are square matrices of the dimension four. They are defined in the standard representation as follows: $\gamma^0 = \beta$, $\gamma^1 = \beta\alpha_x$, $\gamma^2 = \beta\alpha_y$, $\gamma^3 = \beta\alpha_z$, where $\beta = \begin{pmatrix} I & 0 \\ 0 & -I \end{pmatrix}$ and $\vec{\alpha} = \begin{pmatrix} 0 & \vec{\sigma} \\ \vec{\sigma} & 0 \end{pmatrix}$. The matrix I denotes the identity matrix of the dimension two. The matrices $\vec{\sigma}$

© Springer International Publishing AG 2018
I. Tolstikhina et al., *Basic Atomic Interactions of Accelerated Heavy Ions in Matter*, Springer Series on Atomic, Optical, and Plasma Physics 98, https://doi.org/10.1007/978-3-319-74992-1

In this formalism ψ_a means that a particle a has been annihilated (or the anti particle of a has been created) and $\overline{\psi}_a$ means that a particle a has been created (or the anti particle of a has been annihilated). For example the field operator ψ_e can means the annihilation of an electron or the creation of a positron.

The corresponding hamiltonians of the β^- and β^+ decays, H_- and H_+, can be written according to the four point interaction as:

$$H_- = G_F(\overline{\psi}_p \gamma^\mu \psi_n)(\overline{\psi}_e \gamma_\mu \psi_\nu) + h.c. \tag{B.2}$$

and

$$H_+ = G_F(\overline{\psi}_n \gamma^\mu \psi_p)(\overline{\psi}_\nu \gamma_\mu \psi_e) + h.c., \tag{B.3}$$

respectively. The factor G_F is the Fermi coupling constant, $G_F/(\hbar c)^3 = 1.16637 \cdot 10^{-5}$ GeV^{-2}.

B.1   Allowed Transitions and Selection Rules

The lepton current can be aproximated by a plane wave. The Taylor expansion of this plane wave leads to:

$$\psi_e^\dagger \psi_\nu \sim \exp(-i\overrightarrow{p_l} \cdot \overrightarrow{r}) = 1 - i\overrightarrow{p_l} \cdot \overrightarrow{r} + \cdots \tag{B.4}$$

where $\overrightarrow{p_l}$ is the momentum carried by the two leptons. The term $\overrightarrow{p_l} \cdot \overrightarrow{r}$ in the development can be neglected at first order. Variable r corresponds to the radius of the nucleus of a few fermi. A typical momentum carried by a lepton is of the order of a few MeV/c. Taking into account that $\hbar c \sim 197$ MeV fm^{-1} and using the corresponding de Broglie wave length

$$\lambda = \frac{\hbar c}{p_l c}$$

we obtain $p_l r \sim r/\lambda \sim 1/200$. The first order in the expansion (B.4) corresponds to allowed transitions, where the orbital angular momentum carried by the leptons is $\overrightarrow{l} = \overrightarrow{r} \times \overrightarrow{p_l} = 0$. The higher orders ($l \neq 0$) are called l-order forbidden transitions. Since $p_l r$ is of the order of 10^{-2}, and enters squared into the equation for the decay constant, (10.9), the decay constant for each l-order forbidden transition is suppressed by a factor of about 10^4 compared to the $(l-1)$-order.

---

denote the Pauli matrices, basis of the SU(2) group, which are defined as follows: $\sigma_x = \begin{pmatrix} 0 & 1 \\ 1 & 0 \end{pmatrix}$ $\sigma_y = \begin{pmatrix} 0 & -i \\ i & 0 \end{pmatrix}$ and $\sigma_z = \begin{pmatrix} 1 & 0 \\ 0 & -1 \end{pmatrix}$. In addition one defines the matrix $\gamma^5 = i\gamma^0 \gamma^1 \gamma^2 \gamma^3$.

The hadronic part of the hamiltonian has to contain a term accounting for the annihilation of the neutron(proton) and the creation of a proton(neutron). This is realized by introducing an isospin operator τ_+ such that

$$M_{if} = \left\langle f \left| \sum_j \tau_+^{(j)} \right| i \right\rangle \tag{B.5}$$

where the index j runs over all nucleons in the nucleus. The nuclear state can be represented as

$$|\alpha J M \pi\rangle$$

where J is the angular momentum of the nucleus, M is the projection on a selected z-axis, π the parity, and α are all other quantum numbers which are not relevant for the present discussion. Without polarisation of the nuclei, one has to account for all possible M states. One can obtain the reduced matrix element defined by the Wigner-Eckart theorem:

$$\langle \alpha_f J_f \pi_f || T(k, \pi) || \alpha_i J_i \pi_i \rangle \tag{B.6}$$

where $T(k, \pi)$ is an operator of tensorial order k and with parity π. The selection rules for the angular momentum and parity are then:

$$|J_f - J_i| \le k \le J_f + J_i \text{ and } \pi = \pi_f \pi_i \tag{B.7}$$

For $k = 0$, the tensor $T(k, \pi)$ reduces to identity and we obtain the case (B.5) called Fermi transition:

$$M_F = \left\langle f \left| \sum_j \tau_+^{(j)} \right| i \right\rangle \tag{B.8}$$

According to (B.7) the selection rules for such transitions are:

$$\Delta J = 0 \text{ and } \pi_i = \pi_f \tag{B.9}$$

Transitions changing the nuclear momentum by one unit $\Delta J = 1$ have been observed experimentally (e.g. the transition $^6\text{He} \rightarrow {}^6\text{Li} + e^- + \overline{\nu_e}$). For $k = 1$ the tensor $T(k, \pi)$ reduces to the spin operator $\vec{\sigma}$ and the transition corresponding to this matrix element is called Gamow-Teller:

$$M_{GT} = \left\langle f \left| \sum_j \vec{\sigma}^{(j)} \tau_+^{(j)} \right| i \right\rangle \tag{B.10}$$

According to (B.7) the selection rules for this transition are:

$$\Delta J = 1 \quad \text{and} \quad \pi_i = \pi_f \tag{B.11}$$

$$\text{and,}$$

$$\Delta J = 0 \quad \text{and} \quad \pi_i = \pi_f \tag{B.12}$$

$$\text{excluding } J_i^\pi = J_f^\pi = 0^+$$

In general the nuclear matrix element can also include a mixture of the two terms M_F and M_{GT}. There can be three types of transitions:

- the pure Fermi transition ($0^+ \longrightarrow 0^+$);
- the pure Gamow-Teller transition ($\Delta J = 1$);
- and a mixture of these two transition types ($\Delta J = 0$ except ($0^+ \longrightarrow 0^+$)).

The existence of Gamow-Teller terms shows that the nature of the weak interaction is not a pure vector interaction. In order to include other possible types of interactions, the Hamiltonian can be expanded to a more general form:

$$H = G_F \sum_{i=S,V,T,A,P} C_i(\overline{\psi_p} O^i \psi_n)(\psi_e O_i \psi_v) \tag{B.13}$$

where O^i with weights C_i are operators whose properties are characterized by a proper Lorentz transformation. There are 16 linearly independent 4×4 matrices which fall into five classes: Scalar I, Vector γ^μ, Tensor $\sigma^{\mu\nu}$, Axial vector $\gamma^5\gamma^\mu$ and Pseudoscalar γ^5. In the non-relativistic limit, Fermi transitions are of type S or V and Gamow-Teller transitions are of type T or A.

In the 1950s, spectacular experimental and theoretical findings have modified the understanding on the weak interaction. The starting push came from the so called $\theta - \tau$ puzzle. The particles θ and τ, known presently as kaon K^+, have equal masses and mean life-times. For a long time they were considered as two different particles because their decay products have different parities. If the conservation of parity holds in weak interaction then the θ and τ particles must also have different parities. In 1954, T. D. Lee and C. N. Yang suggested that conservation of parity might be broken in weak interaction and proposed several experiments for testing this [532]. If parity is conserved then the non-vanishing observables are scalar quantities such as life time, angular correlations $\overrightarrow{p_1} \cdot \overrightarrow{p_2}$, spin correlations $\overrightarrow{S_1} \cdot \overrightarrow{S_2}$, etc. In the case of parity violation, one can in addition observe pseudo-scalar quantities and one has to add to the hamiltonian (B.13) a pseudoscalar term H':

$$H' = G \sum_{i=S,V,T,A,P} C'_i(\overline{\psi_p} O^i \psi_n)(\overline{\psi_e} O_i \gamma^5 \psi_v) \tag{B.14}$$

where C'_i are the weights of the pseudo-scalar operators and they express the strength of parity violation. The weak interaction Hamiltonian (B.13) is then generalized to:

$$H = G \sum_{i=S,V,T,A,P} (\overline{\psi_p} O^i \psi_n)[\overline{\psi_e} O_i (C_i + C_i' \gamma^5) \psi_v] \tag{B.15}$$

In 1957, C. S. Wu performed the first experiment showing the evidence for non-conservation of parity in β-decay [532]. Direction of the electron emission in Gamow-Teller transition of polarized ^{60}Co has been investigated. From the conservation of angular momentum, the spins of the emitted electron and antineutrino have to be parallel to the spin of the parent nucleus. If parity is not conserved, one expects a non-zero pseudo-scarlar observable such as $\langle \vec{S} \cdot \vec{p} \rangle \neq 0$, which would mean that there is a preferred direction of the momentum. The experiment realised by C. S. Wu showed that electrons are emitted predominantly in the opposite direction of their spin, that is, they have a negative helicity. Further complementary experiments have been performed [533]. It has been shown that the weak interaction acts always on the left-handed particle component or on the right-handed antiparticle component: the parity violation is maximal. Therefore, one has to project the wave functions of particles on left chirality state, and antiparticles on the right chirality state:

$$a\psi = \frac{1 - \gamma^5}{2} \psi \tag{B.16}$$

and

$$\overline{a\psi} = \overline{\psi} \frac{1 + \gamma^5}{2} \tag{B.17}$$

where the operators a and $\overline{a}$ denote the projection operators on the left handed and right handed chirality states, respectively.

The restriction to the experimentally observed helicity components and the Lorentz invariance of the β-decay Hamiltonian exclude all coupling except V and A. This has led to the construction of the $V - A$ theory [534]. The Hamiltonians (B.2) and (B.3) of the β^- and β^+ decays can be rewritten as:

$$H_- = \frac{G_F}{\sqrt{2}} (\overline{\psi_p} \gamma^\mu (C_V + C_A \gamma^5) \psi_n)(\overline{a\psi_e} \gamma_\mu a\psi_v) \tag{B.18}$$

and

$$H_+ = \frac{G_F}{\sqrt{2}} (\overline{\psi_n} \gamma^\mu (C_V + C_A \gamma^5) \psi_p)(\overline{a\psi_v} \gamma_\mu a\psi_e) \tag{B.19}$$

respectively. Here C_V and C_A indicate the coupling constants of the nuclear part of the Hamiltonian. The relation between these constants has been determined and equals to $C_A/C_V = -1.255$ (6). Hamiltonians written in form (B.18) and (B.19) include the violation of parity.

Appendix C
Theoretical Description of the 3/2 EC-Decay Rate Ratio Found in H- and He-Like Stored Ions

In Sect. 10.4, we consider an example of measurement, showing that the decay rate can depend on the charge state of the ion. For the H-like and He-like case, it is also possible to find the 3/2 result [535] by taking into account the well defined helicity of the neutrino. In this approach that we describe in the following, we consider only the transitions from ground state to ground state. The difference between the following approach is that the wave function of the neutrino is constructed as beeing a left-handed particle while in [529] this is the operator H which projects the particle into its left chirality state. An expression of the EC probabilities for H- and He-like ions are found in the next two sections.

C.1 H-Like Case

In the H-like case, the spin of the nucleus in the initial state is $I_i = 1$ and the spin of the K-electron is $S_e = 1/2$. This leads for the initial state to a total angular momentum of $F_i = 1 \pm 1/2$, that is, $F_i = 1/2$ or $F_i = 3/2$ hyperfine states. In the final state however, after a pure Gamow-Teller transition occured, the nuclear spin is $I_f = 0$. The neutrino has a spin $S_\nu = 1/2$ and does not carry any angular momentum in such an allowed transition, i.e. $L_\nu = 0$. Thus, the total angular momentum of the system in the final state is restricted to the state $F_f = 1/2$. It follows that from the possible transitions given by:

$$F_i = 3/2 \nrightarrow F_f = 1/2$$
$$F_i = 1/2 \rightarrow F_f = 1/2$$

only the latter one is permitted. This means that only transitions from the $F_i = 1/2$ hyperfine state can contribute to the decay. Furthermore, the mother nucleus having a positive magnetic moment [536], the electrons are populated in the lower energy state which coresponds to $F_i = 1/2$, the possible populated upper hyperfine states beeing

© Springer International Publishing AG 2018
I. Tolstikhina et al., *Basic Atomic Interactions of Accelerated Heavy Ions in Matter*, Springer Series on Atomic, Optical, and Plasma Physics 98,
https://doi.org/10.1007/978-3-319-74992-1

de-excited in lower hyperfine states with a relaxation time much shorter than the cooling time. Thus there are two initial states corresponding to the spin projections ($|F = 1/2, m_F = +1/2\rangle \equiv |+\rangle_{Ne}$ and $|F = 1/2, m_F = -1/2\rangle \equiv |-\rangle_{Ne}$) of equal occupation probabilities $P_i^H(m_F) = 1/2$ with $m_F = 1/2$ or $m_F = -1/2$, m_F being the eigenvalue of the spin projection operator. In the final state, there are two accessible states: $|0, 0\rangle_N \otimes |1/2, +1/2\rangle_v$ and $|0, 0\rangle_N \otimes |1/2, -1/2\rangle_v$. In addition, the neutrino has a well defined helicity ($h = -1$). Let assume that the neutrino is emmitted with a momentum $\vec{p} = |p|\vec{n}$, where $\vec{n}$ is the unit vector along the direction of the neutrino emission and is defined as $\vec{n} = \begin{pmatrix} \sin\theta\cos\phi \\ \sin\theta\sin\phi \\ \cos\theta \end{pmatrix}$.

If $\vec{n}$ is the quantization axis, then the projection of the neutrino spin can be only $-1/2$ due to its negative helicity. In this frame the final state is expressed as follow:

$$|0, 0\rangle_N \otimes |1/2, -1/2\rangle'_v \equiv |-\rangle'_{Nv} \tag{C.1}$$

where the sign $'$ denotes a basis with the quantization axis along the direction of the neutrino emission $\vec{n}$. The total orbital electron capture probability is thus:

$$P_H = \frac{1}{2}|_{Nv}'\langle -|H|+\rangle_{Ne}|^2 + \frac{1}{2}|_{Nv}'\langle -|H|-\rangle_{Ne}|^2 \tag{C.2}$$

The H-like and He-like ions are unpolarized in the ESR. This means that the quantization axis z of the laboratory frame is chosen arbitrarily. Therefore we can choose the quantization axis $\vec{n}$ for the initial states as well. Note that this is not appropriate for the case of polarized H-like ions, although the ratio stays unchanged with respect to the unpolarized one (see Appendix). The total probability becomes:

$$P_H = \frac{1}{2}|_{Nv}'\langle -|H|+\rangle'_{Ne}|^2 + \frac{1}{2}|_{Nv}'\langle -|H|-\rangle'_{Ne}|^2 \tag{C.3}$$

Since the Hamiltonian has nonzero matrix elements only between states with identical total angular momentum and its projection, the later relation reduces to:

$$P_H = \frac{1}{2}|_{Nv}'\langle -|H|-\rangle'_{Ne}|^2 \equiv A/2 \tag{C.4}$$

where we have defined $A = |_{Nv}'\langle -|H|-\rangle'_{Ne}|^2$. Finally we find,

$$2P_H = A \tag{C.5}$$

C.2 He-Like Case

For helium-like ^{142}Pm ions one uses similar arguments.

C.2.1 Initial States

In the initial state, the two electrons of the K shell are antiparallel due to the Pauli exclusion principle. Therefore they are in a singlet state:

$$|0, 0\rangle_{e_1 e_2} = \frac{|+\rangle_{e_1}|-\rangle_{e_2} - |-\rangle_{e_1}|+\rangle_{e_2}}{\sqrt{2}} \tag{C.6}$$

In the initial state, the nuclear spin $I = 1$ has $2I_i + 1 = 3$ projections, corresponding to

$$|1, M\rangle_N = \begin{cases} |1, +1\rangle_N \\ |1, \ \ 0\rangle_N \\ |1, -1\rangle_N \end{cases} \tag{C.7}$$

Finally, the initial state is the tensor product of the nuclear and leptonic part:

$$|\Psi_i\rangle = |1, M\rangle_N \otimes \frac{|+\rangle_{e_1}|-\rangle_{e_2} - |-\rangle_{e_1}|+\rangle_{e_2}}{\sqrt{2}} \tag{C.8}$$

C.2.2 Final States

In the final state, the nuclear spin is $I = 0$, the remaining electron has spin $S_e = 1/2$ and the neutrino has spin $S_\nu = 1/2$. Because of the conservation of the total angular momentum in the EC-decay, the spin of the remaining electron and the spin of the neutrino have to couple to $F = 1$. Again, if we choose the quantization axis $\vec{n}$ for the projection of the spin of the neutrino, we obtain two final states:

$$\begin{cases} |0, 0\rangle_N \otimes |+\rangle_e \otimes |-\rangle'_\nu \\ |0, 0\rangle_N \otimes |-\rangle_e \otimes |-\rangle'_\nu \end{cases} \tag{C.9}$$

It is clear that in the "neutrino frame", the projection of the total angular momentum of the final state can never be $M' = 1$ because of the well defined helicity of the neutrino. However the projection can be 0 or -1 depending on the projection of the remaining electron.

C.2.3 Total EC-Probability

The total orbital electron capture probability can be expressed as follow:

$$P_{He} = \sum_M \frac{1}{2F_i + 1} P_M \qquad (C.10)$$

where $F_i = 1$ is the total angular momentum with projection $M = 1, 0, -1$, and the probability P_M is:

$$
\begin{aligned}
P_M = 2 \times [\ &|_N\langle 0,0|_e\langle -|_v{}^\backprime\langle -|H|1,M\rangle_N\{|+\rangle_{e_1}|-\rangle_{e_2} - |-\rangle_{e_1}|+\rangle_{e_2}\}/\sqrt{2}|^2 \\
+\ &|_N\langle 0,0|_e\langle +|_v{}^\backprime\langle -|H|1,M\rangle_N\{|+\rangle_{e_1}|-\rangle_{e_2} - |-\rangle_{e_1}|+\rangle_{e_2}\}/\sqrt{2}|^2\ ]
\end{aligned} \qquad (C.11)
$$

The factor 2 express the fact that there are two possibilities of capturing an electron, that is, the electron e_1 or the electron e_2. Let us first assume that the electron e_2 is captured. In this case, the remaining electron e_1 does not participate on the weak decay, and we have $_{e_1}\langle m|H|m'\rangle_{e_1} = {}_{e_1}\langle m|m'\rangle_{e_1} = \delta_{m,m'}$, where $m, m' = \pm 1/2$. Therefore equation (C.11) can be simplified to

$$
\begin{aligned}
P_M = &|_N\langle 0,0|_v{}^\backprime\langle -|H|1,M\rangle_N|+\rangle_{e_2}|^2 \\
+ &|_N\langle 0,0|_v{}^\backprime\langle -|H|1,M\rangle_N|-\rangle_{e_2}|^2
\end{aligned} \qquad (C.12)
$$

The hamiltonian H having non zero matrix elements only between states with same angular momentum and projections, it can be shown [528] using the proper Clebsch-Gorden coefficients that

$$
\begin{aligned}
P_{+1} &= 0 \\
P_0 &= A/3 \\
P_{-1} &= 2A/3
\end{aligned} \qquad (C.13)
$$

The total orbital electron capture probability given by equation (C.10) becomes:

$$P_{He} = \frac{1}{3}P_{+1} + \frac{1}{3}P_0 + \frac{1}{3}P_{-1} = \frac{1}{3} \times (0 + \frac{1}{3}A + \frac{2}{3}A) \qquad (C.14)$$

or

$$3P_{He} = A \qquad (C.15)$$

Using the latter relation (C.15) and the relation (C.5), found in H-like case, we find finally:

$$3P_{He} = A = 2P_H \Leftrightarrow \frac{P_H}{P_{He}} = \frac{3}{2} \qquad (C.16)$$

This ratio of $3/2$ is in good agreement with the measured ones. One can show that this result is the same in case of polarized H- and He-like ^{142}Pm ions. The ratio $P_H/P_{He} = 3/2$ is valid for all EC transitions $1^+ \rightarrow 0^+$.

References

1. H.D. Betz, Charge states and charge-changing cross sections of fast heavy ions penetrating through gaseous and solid media. Rev. Mod. Phys. **44**, 465–539 (1972)
2. I.A. Sellin (ed.), *Structure and Collisions of Ions and Atoms* (Springer, Berlin, 1978)
3. S.P. Ahlen, Theoretical and experimental aspects of the energy loss of relativistic heavily ionizing particles. Rev. Mod. Phys. **52**, 121–173 (1980)
4. S. Datz, H.D. Betz, *Applied Atomic Collisions Physics*, vol. 4. Condensed Matter: Heavy Ion Charge States, (Academic Press, London, 1983), pp. 1–38
5. J.F. Ziegler, J.P. Biersack, U. Littmark, in *The Stopping and Range of Ions in Solids*, Vol. 1. (Pergamon, New York, 1985) (new edition in 1996)
6. N. Stolterfoht, R.D. DuBois, R.D. Rivarola, *Electron Emission in Heavy Ion-Atom Collisions* (Springer, Berlin, 1997)
7. U. Amaldi, G. Kraft, Recent applications of synchrotrons in cancer therapy with carbon ions. Europhys. News **36**, 114–118 (2005)
8. D. Schardt, D. Th Elsässer, Schulz-Ertner, Heavy-ion tumor therapy: physical and radiobiological benefits. Rev. Mod. Phys. **82**, 383–425 (2010)
9. C. Trautmann, Micro- nano-engineering with ion tracks, in *Ion Beams in Nanoscience and Technology*, ed. by R. Hellborg, H.J. Whitlow, Y. Zhang (Springer, Berlin, 2009)
10. D. Schulz-Ertner, A. Nikoghosyan, Ch. Thilmann, O. Th Haberer, Ch. Jäke, M. Karger, G. Scholz, M. Kraft, J.Debus Wannenmacher, Carbon ion radiotherapy for chordomas and low-grade chondrosarcomas of the skull base. Int. J. Radiat. Onc. Biol. Phys. **58**, 631–640 (2004)
11. V.P. Shevelko, H. Tawara (eds.), *Atomic Processes in Basic and Applied Physics* (Springer, Berlin, 2012)
12. B. Schmidt, K. Wetzig, *Ion Beams in Materials Processing and Analysis* (Springer, Berlin, 2013)
13. H. Tsujii, T. Kamada, T. Shirai, K. Noda, H. Tsuji, K. Karasawa (eds.), *Carbon-Ion Radiotherapy. Principles, Practices, and Treatment Planning* (Springer, Tokyo, 2014)
14. P. Sigmund, *Particle Penetration and Radiation Effects*, vol. 2 (Springer, Berlin, 2014)
15. Yu.Ts. Oganessian, V.K. Utyonkov, Super-heavy element research. Rep. Prog. Phys. **78**, 036301 (22pp) (2015)
16. H.A. Bethe, E.E. Salpeter, *Quantum Mechanics of One- and Two-Electron Atoms* (Springer, Berlin, 1957)
17. L. Landau, E. Lifshitz, *Quantum Mechanics: Non-Relativistic Theory*, vol. 3, 3rd edn. (Pergamon Press, Oxford, 1977)

© Springer International Publishing AG 2018
I. Tolstikhina et al., *Basic Atomic Interactions of Accelerated Heavy Ions in Matter*, Springer Series on Atomic, Optical, and Plasma Physics 98, https://doi.org/10.1007/978-3-319-74992-1

18. D.R. Bates, *Atomic and Molecular Processes* (Academic Press, New York, 1962)
19. H.A. Bethe, *Intermediate Quantum Mechanics* (W.A. Benjamin, New York, 1964)
20. N.F. Mott, H.S.W. Massey, *The Theory of Atomic Collisions* (Clarendon Press, Oxford, 1965)
21. L.P. Presnyakov, X-ray spectroscopy of high-temperature plasma. Sov. Phys. Usp. **19**, 387–399 (1976). [Usp. Fiz. Nauk, **119**, 49–73 (1976)]
22. I.I. Sobelman, L.A. Vainshtein, E.A. Yukov, *Excitation of Atoms and Broadening of Spectral Lines* (Springer, Berlin, 1981)
23. B.M. Smirnov, Ionization in low-energy atomic collisions. Sov. Phys. Usp. **24**, 251–275 (1981). [Usp. Fiz. Nauk **133**, 569–616 (1981)]
24. R.K. Janev, L.P. Presnyakov, Collision processes of multiply charged ions with atoms. Phys. Rep. **70**, 1–107 (1981)
25. R.K. Janev, L.P. Presnyakov, V.P. Shevelko, *Physics of Highly Charged Ions* (Springer, Berlin, 1985)
26. R.K. Janev, W.D. Langer, K.J. Evans, D.E.J. Post, *Elementary Processes in Hydrogen-Helium Plasmas* (Springer, Berlin, 1987)
27. R.K. Janev, D. Reiter, U. Samm, Collision Processes in Low-Temperature Hydrogen Plasmas. Berichte des Forschungszentrums Julich:4105, ISSN 0944-2952. Institut fur Plasmaphysik EURATOM Association, Mitglied im Trilateral Euregio Cluster Ju1-4105, pp. 1–188 (2005)
28. E.A. Solov'ev, Nonadiabatic transitions in atomic collisions. Sov. Phys. Uspekhi **32**(3), 228–250 (1989). [UFN **157**, 437-476 (1989)]
29. W. Fritsch, C.D. Lin, The semiclassical close-coupling description of atomic collisions: recent developments and results. Phys. Rep. **202**, 1–97 (1991)
30. B.H. Bransden, M.R.C. McDowell, *Charge Exchange and the Theory of Ion-Atom Collisions* (Clarendon Press, Oxfors, 1992)
31. V.P. Shevelko, L.A. Vainshtein, *Atomic Physics for Hot Plasmas* (IOP Publ, Bristol, 1993)
32. R.K. Janev (ed.), *Atomic and Molecular Processes in Fusion Edge Plasmas* (Plenum Press, New York, 1995)
33. V.P. Shevelko, H. Tawara, *Atomic Multielectron Processes* (Springer, Berlin, 1998)
34. V.S. Lebedev, I.L. Beigman, *Physics of Highly Excited Atoms and Ions* (Springer, Berlin, 1998)
35. H.F. Beyer, V.P. Shevelko (eds.), *Atomic Physics with Heavy Ions* (Springer, Berlin, 1999)
36. B.M. Smirnov, Atomic structure and the resonant charge exchange process. Phys. Usp. **44**, 221–253 (2001). [Usp. Fiz. Nauk **171**, 233–266 (2001)]
37. H.F. Beyer, V.P. Shevelko, *Introduction to the Physics of Highly Charged Ions* (IOP Publ, Bristol, 2003)
38. F.J. Currell (ed.), *The Physics of Multiply and Highly Charged Ions* (Kluwer Acad. Publ, Dordrecht, 2003)
39. B.Y. Sharkov, Overview of Russian heavy-ion inertial fusion energy program. Nucl. Instrum. Meth. Phys. Res. A **277**, 14–20 (2007)
40. J. Eichler, T. Stöhlker, Radiative electron capture in relativistic ionatom collisions and the photoelectric effect in hydrogen-like high-Z systems. Phys. Rep. **439**, 1–99 (2007)
41. V. Shevelko, H. Tawara (eds.), *Atomic Processes in Basic and Applied Physics* (Springer, Heidelberg, 2012)
42. I. Tolstikhina, V.P. Shevelko, Collision processes involving heavy many-electron ions interacting with neutral atoms. Phys. Usp. **56**, 213–242 (2013). [Usp. Fiz. Nauk **183**, 225–255 (2013)]
43. I. Tolstikhina, V.P. Shevelko, Influence of atomic processes on charge states and fractions of fast heavy ions passing through gaseous, solid and plasma targets (accepted for publicatopn 2017); Phys. Usp. (2017) https://doi.org/10.3367/UFNr.2017.02.038071
44. B.G. Logan, L.J. Perkins, J.J. Barnard, Direct drive heavy-ion-beam inertial fusion at high coupling efficiency. Phy. Plasmas **15**, 072701–7 (2008)
45. H. Geissel, H. Weick, C. Scheidenberger, R. Bimbot, D. Gardés, Experimental studies of heavy-ion slowing down in matter. Nucl. Instrum. Meth. Phys. Res., B **195**, 3–54 (2002)

46. H. Geissel, G. Munzenberg, K. Riisager, Secondary exotic nuclear beams. Ann. Rev. Nucl. Part. Sci. **45**, 163–203 (1995)
47. V.E. Fortov, D.H.H. Hoffmann, B.Y. Sharkov, Intense ion beams for generating extreme states of matter. PhysicsUspekhi **51**, 109131 (2008). [Usp. Fiz. Nauk **178**, 113–138 (2008)]
48. K. Dennerl, Charge transfer reactions. Space Sci. Rev. **157**, 57–91 (2010)
49. H. Wiedemann, *Particle Accelerator Physics* (Springer, Berlin, 2007)
50. https://www-amdis.iaea.org/
51. https://www.gsi.de/forschungbeschleuniger/fair.htm
52. http://nica.jinr.ru/ru/
53. W. Wien, K. Sitzungsber, Preuss. Akad. Wiss. S.773786 (1911)
54. Y. Yano, The RIKEN RI Beam Factory Project: a status report. Nucl. Instrum. Methods Phys. Res., Sect. B **261–270**, 1009–1013 (2007)
55. https://frib.msu.edu/
56. J. Yang, J. Xia, G. Xiao, H. Xu, G. Zhao, X. Zhou, X. Ma, Y. He, L. Ma, D. Gao, High intensity heavy ion accelerator facility (HIAF) in China. Nucl. Instrum. Methods Phys. Res., Sect. B **317**, 263–265 (2013)
57. H. Folger (ed.), Heavy ion targets and related phenomena (special issue). Nucl. Instr. Meth. A **282** (1989)
58. A. David, P. Adams, B. Ryan, *Prediction and Characterization of Heat-Affected Zone Formation due to Neighboring Nickel-Aluminum Multilayer Foil Reaction (Report SAND 2015–8325* (SANDIA, USA, 2015)
59. S.G. Lebedev, A.S. Lebedev, Calculation of the lifetimes of thin stripper targets under bombardment of intense pulsed ions. Phys. Rev. ST **11**, 020401–7 (2008)
60. N.A. Tahir, V. Kim, B. Schlitt, W. Barth, L. Groening, I.V. Lomonosov, A.R. Piriz, Th. Stöhlker, H. Vormann, Three-dimensional thermal simulations of thin solid carbon foils for charge stripping of high current uranium ion beams at the proposed new GSI heavy ion linac. Phys. Rev. ST Accel. Beams **17**, 041003 (2014)
61. N.A. Tahir, F. Burkart, A. Shutov, R. Schmidt, D. Wollmann, A.R. Piriz, Simulations of beam-matter interaction experiments at the CERN HiRadMat facility and prospects of high-energy-density physics research. Phys. Rev E **90**, 063112–10 (2014)
62. Y. Geissel, Y. Laichter, W.F.W. Schneider, P. Armbruster, Energy loss and energy loss straggling of fast heavy ions in matter. Nucl. Instr. Meth. **194**, 21–29 (1982)
63. R. Bimbot, C. Cabot, D. Gardes, H. Gauven, I. Orliange, L. de Reilhac, K. Subotic, F. Hubert, Stopping power of gases for heavy ions gas-solid effect: II. 2–6 MeV/u Cu, Kr and Ag projectiles. Nucl. Instrum. Methods B **44**, 19–34 (1989)
64. W. Barth, A. Adonin, C.E. Düllmann, M. Heilmann, R. Hollinger, E. Jäger, J. Khuyagbaatar, J. Krier, P. Scharrer, H. Vormann, A. Yakushev, U^{28+}-intensity record applying a H2-gas stripper cell. Phys. Rev. Special Topics—Acce; erators and Beams **18**, 040101–9 (2015)
65. P. Scharrer, C.E. Düllmann, W. Barth, J. Khuyagbaatar, A. Yakushev, M. Bevcic, P. Gerhard, L. Groening, K.P. Horn, E. Jäger, J. Krier, H. Vormann, Measurements of charge state distributions of 0.74 and 1.4 MeV/u heavy ions passing through dilute gases. Phys. Rev. Accelerators Beams **20**, 043503–14 (2017)
66. D. Gardés, M. Chabot, V. Nectoux, G. Maynard, C. Deutsch, I. Roudskoi, Experimental study of stopping power for high Z ion in hydrogen. Nucl. Instr. Meth. A **415**, 698–702 (1998)
67. D.H.H. Hoffmann, K. Weyrich, H. Wahl, D. Gardés, R. Bimbot, C. Fleurier, Energy loss of heavy ions in a plasma target. Phys. Rev. A **42**, 2313–2321 (1990)
68. V. Mintsev, V. Gryaznov, M. Kulish, V.E. Fortov, B. Sharkov, A. Golubev, A. Fertman, N. Mescheryakov, W. Su, D.H.H. Hoffmann, M. Stetter, R. Bock, M. Roth, C. Stockl, D. Gardés, On measurements of stopping power in explosively driven plasma targets. Nucl. Instr. Meth. A **415**, 715–719 (1998)
69. M. Ogawa, Y. Oguri, U. Neuner, K. Nishigori, A. Sakumi, K. Shibata, J. Kobayashi, M. Kojima, M. Yoshida, J. Hasegawa, Laser heated dE/dX experiments in Japan. Nucl. Instr. Meth. A **464**, 72–79 (2001)

70. M. Chabot, D. Gardés, P. Box, J. Kiener, C. Deutsch, G. Maynard, M. André, C. Fleurier, D. Hong, K. Wohrer, Ion stopping in dense plasma target for high energy density physics. Phys. Rev. E **51**, 3504–10 (1995)

71. G. Rouille, M. Bassan, C. Commeaux, J.P. Didelez, C. Schaerf, V. Bellini, J.P. Bocquet, M. Capogni, M. Castoldi, A.D. Angelo, G. Gervino, F. Ghio, B. Girolami, M. Guidal, E. Hourany, R. Kunne, P.L. Sandri, A. Lleres, D. Moricciani, D. Rebreyend, A. Zucchiatti, A polarized HD target factory in Europe. Nucl. Instr. Meth. A **464**, 428–432 (2001)

72. C. Fleurier, A. Sanba, D. Hong, J. Mathias, J.C. Pellicer, Plasma diagnostics in the heavy ion beam-dense plasma interaction experiment at Orsay. J. de Phys. C **7**, 141–149 (1988)

73. S. Eisenbarth, O. Rosmej, V.P. Shevelko, A. Blazevich, D.H.H. Hoffmann, Numerical simulations of the projectile ion charge difference in sold and gaseous stopping matter. Laser and Particles Beams **27**, 601–611 (2007)

74. Th Peter, J. Meyer-ter-Vehn, Energy loss of heavy ions in palsma: I. Linear and nonlinear Vlasov theory for the stopping power. Rev. A **43**, 1998–2014 (1991)

75. Th Peter, J. Meyer-ter-Vehn, Energy loss of heavy ions in palsma: II. Nonequilibrium charge states and stopping powers. Rev. A **43**, 2015–2030 (1991)

76. J. Jacoby, D.H.H. Hoffmann, W. Laux, R.W. Miiller, H. Wahl, K. Weyrich, E. Boggasch, B. Heimrich, C. Stockl, H. Wetzler, S. Miyamoto, Stopping of heavy ions in a hydrogen plasma. Phys. Rev. Lett. **74**, 1550–1553 (1995)

77. D. Schulz-Ertner, H. Tsujii, Particle radiation therapy using proton and heavier ion beams. J. Clin. Oncol. **25**, 953–964 (2007)

78. D. Ohsawa, Y. Sato, Y. Okada, V.P. Shevelko, F. Soga, 6.0-10.0-MeV/u He2+-ion-induced electron emission from water vapor. Phys. Rev. A **72**, 062710–13 (2005)

79. O.I. Obolensky, E. Surdutovich, I. Pshenichnov, I. Mishustin, A.V. Solovyov, W. Greiner, Ion beam cancer therapy: fundamental aspects of the problem. Nucl. Instrum. Methods B **266**, 1623–1628 (2008)

80. M.A. Quinto, J.M. Monti, P.D. Montenegro, O.A. Fojón, Ch. Champion, R.D. Rivarola, Single ionization and capture cross sections from biological molecules by bare projectile impact. Eur. Phys. J. B **71**(35), 1–6 (2017)

81. C. Scheidenberger, Untersuchung der Abbremsung relativistischer Schwerionen in Materie im Energiebereich 100–1000 MeV/u. Ph.D. thesis, Giessen Physikalische Institut, Giessen 1994

82. A. Närmann, P. Sigmund, Statistics of energy loss and charge exchange of penetrating particles: higher moments and transients. Phys. Rev. A **49**, 4709–4715 (1994)

83. L. Landau, On the energy loss of fast particles by ionization. J. Phys. USSR **8**, 201–205 (1944)

84. P.V. Vavilov, Ionizotion losses oi high-energy heavy particles sov. Phys. JETP **5**, 749–751 (1957)

85. C. Scheidenberger, H. Geissel, Penetration of relativistic heavy ions through matter. Nucl. Instrum. Meth. B **135**, 25–34 (1998)

86. H.H. Andersen, P. Sigmund (eds.): Proceedings of the International Workshop on the Slowing Down of Ions STOPOI. Nucl. 69 (1992)

87. H.A. Bethe, Moliere's theory of multiple scattering. Phys. Rev. **89**, 1256–1266 (1953)

88. V.L. Highland, Some practical remarks on multiple scattering. Nucl. Instrum. Methods **129**, 497–499 (1975)

89. Particle Data Group, S. Eidelman, K.G. Hayes, K.A. Olive, M. Aguilar-Benitez, C. Amsler, D. Asner, K.C. Babu, R.M. Barnett, J. Beringer, P.R. Burchat, C.D. Carone, S. Caso, G. Conforto, D. Dahl, G. DÁmbrosio, M. Doser, J.L. Feng, T. Gherghetta, L. Gibbons, M. Goodman, et al., Rev. Part. Phy. Phys. Lett. B **592**, 1–5 (2004)

90. G.R. Lynch, O. Dahl, Approximations to multiple Coulomb scattering. Nucl. Instrum. Meth. B **58**, 6–10 (1991)

91. http://www.kayelaby.npl.co.uk/atomic_and_nuclear_physics/4_5/4_5_1.html

92. https://www.nist.gov/pml/stopping-power-range-tables-electrons-protons-and-helium-ions

93. H.H. Lo, W.L. Fite, Electron capture and loss cross sections for fast, heavy particles passing through gases. At. Data **1**, 305–328 (1970)

94. R.C. Dehmel, H.K. Chan, H.H. Fleischmann, Experimental stripping cross sections for atoms and ions in gases, 1950–1970, At. Data **5**, 231–289 (1973)

95. W. Erb, Umladung schwerer Ionen Durchgang Gase und Festköpper im Energiebereich 0,2 bis 1,4 MeV/u. GSI Report P-78, pp. 1–113, GSI, Darmstadt 1978

96. B. Franzke, Vacuum requirements for heavy ion synchrotrons. IEEE Trans. Nucl. Sci. **28**, 2116–2118 (1981)

97. T. Stöhlker, Ch. Kozhuharov, P.H. Mokler, R.E. Olson, Z. Stachura, A. Warczak, Single and double electron capture in collisions of highly ionized, decelerated Ge ions with Ne. J. Phys. B **25**, 4527–4532 (1992)

98. X. Ma, T. Stöhlker, F. Bosch, O. Brinzanescu, S. Fritzsche, C. Kozhuharov, T. Ludziejewski, P.H. Mokler, Z. Stachura, A. Warczak, State-selective electron capture into He-like U^{90+} ions in collisions with gaseous targets. Phys. Rev. A **64**, 2704–8 (2004)

99. G. Weber, C. Omet, R.D. DuBois, O. de Lucio, T. Stöhlker, C. Brandau, A. Gumberidze, S. Hagmann, S. Hess, C. Kozhuharov, R. Reuschl, P. Spiller, U. Spillmann, N. Steck, M. Thomason, S. Trotsenko, Beam lifetimes and ionization cross sections of U^{28+}. Phys. Rev. ST **12**, 084201–5 (2009)

100. G. Weber, M.O. Herdrich, R.D. DuBois, P.-M. Hillenbrand, H. Beyer, L. Bozyk, T. Gassner, R.E. Grisenti, S. Hagmann, Y.A. Litvinov, F. Nolden, N. Petridis, M.S. Sanjari, D.F.A. Winters, T. Stöhlker, Total projectile electron loss cross sections of U^{28+} ions in collisions with gaseous targets ranging from hydrogen to krypton. Phys. Rev. ST **18**, 034403–8 (2015)

101. G. Weber, M.O. Herdrich, R.D. DuBois, P.-M. Hillenbrand, H. Beyer, L. Bozyk, T. Gassner, R.E. Grisenti, S. Hagmann, Y.A. Litvinov, F. Nolden, N. Petridis, M.S. Sanjari, D.F.A. Winters, Th Stöhlker, Total projectile electron loss cross sections of U^{28+} ions in collisions with gaseous targets ranging from hydrogen to krypton. Phys. Rev. ST **18**, 034403–8 (2015)

102. R.E. Olson, R.L. Watson, V. Horvat, A.N. Peruma, Y. Peng, T. Stöhlker, Projectile electron loss and capture in MeV/u collisions of U^{28+} with H_2, N_2, and Ar. J. Phys. B: At. Mol. Opt. Phys. **37**, 4539–4550 (2004)

103. A.N. Perumal, V. Horvat, R.L. Watson, Y. Peng, K.S. Fruchey, Cross sections for charge change in argon and equilibrium charge states of 3.5 MeV/amu uranium ions passing through argon and carbon targets. Nucl. Instrum. Methods Phy. Res. Sect. B **227**, 251–260 (2005)

104. D. Mueller, L. Grisham, I. Kaganovich, R.L. Watson, V. Horvat, K.E. Zaharakas, M.S. Armel, Multiple electron stripping of 3.4 MeVamu Kr7 and Xe11 in nitrogen. Phys. Plasmas **8**, 1753–1756 (2001)

105. J. Alonso, H. Gould, Charge-changing cross sections for Pb and Xe at velocities up to 4×10^9 cm/s. Phys. Rev. A **26**, 1134–1137 (1982)

106. H. Kuboki, H. Okuno, S. Yokouchi, H. Hasebe, T. Kishida, N. Fukunishi, O. Kamigaito, A. Goto, M. Kase, Y. Yano, Charge-state distribution measurements of ^{238}U and ^{136}Xe at 11 MeV/nucleon using gas charge stripper. Phys. Rev. ST **13**, 093501–9 (2010)

107. H. Okuno, N. Fukunishi, A. Goto, H. Hasebe, H. Imao, O. Kamigaito, M. Kase, H. Kuboki, Y. Yano, S. Yokouchi, Low-Z gas stripper as an alternative to carbon foils for the acceleration of high-power uranium beams. Phys. Rev. ST **14**, 033503–8 (2011)

108. H. Imao, H. Okuno, H. Kuboki, S. Yokouchi, N. Fukunishi, O. Kamigaito, H. Hasebe, T. Watanabe, Y. Watanabe, M. Kase, Y. Yano, Charge stripping of ^{238}U ion beam by helium gas stripper. Phys. Rev. ST **15**, 123501–9 (2012)

109. R.D. DuBois, A.C.F. Santos, R.E. Olson, T. Stöhlker, F. Bosch, A. Brauning-Demian, A. Gumberidze, S. Hagmann, C. Kozhuharov, R. Mann, O. Muthig, U. Spillmann, S. Tachenov, W. Barth, L. Dahl, B. Franzke, J. Glatz, L. Groning, S. Richter, D. Wilms, A. Krämer, K. Ullmann, O. Jagutzki, Electron loss from 0.74- and 1.4-MeV/u low-charge-state argon and xenon ions colliding with neon, nitrogen, and argon. Phys. Rev. A **68**, 042701–8 (2003)

110. A.S.F. Santos, R. DuBois, Scaling laws for single and multiple electron loss from projectiles in collisions with a many-electron target. Phys. Rev. A **69**, 042709–11 (2004)

111. A.B. Voitkiv, J. Ullrich, *Relativistic Collisions of Structured Atomic Particles* (Sprinegr, Berlin, 2008)
112. N.V. Novikov, Y. A. Teplova, Charge-changing cross sections in ion-atom collisions. Bibliography and data. http://cdfe.sinp.msu.ru/services/cccs/HTM/authors.htm
113. V.P. Shevelko, I.L. Beigman, M.S. Litsarev, H. Tawara, IYu. Tolstikhina, G. Weber, Charge-changing processes in collisions of heavy many-electron ions with neutral atoms. Nucl. Instrum. Meth. Phy. Res. B **269**, 1455–1463 (2011)
114. H.A. Kramers, On the theory of X-ray absorption and of the continuous X-ray spectrum. Phil. Mag. **46**, 836–871 (1923)
115. H. Geissel, Untersuchungen zur Abbremsung von Schwerionen in Materie im Energiebereich von (0.5–10) MeV/u. GSI-Report 82–12, Darmstadt, (1982) (in German)
116. H. Geissel, C. Scheidenberger, Slowing down of relativistic heavy ions and new applications. Nucl. Instrum. Meth. Phys. Res., B **136**, 114–124 (1998)
117. C. Scheidenberger, T. Stöhlker, W.E. Meyerhof, H. Geissel, P.H. Mokler, B. Blank, Charge states of relativistic heavy ions in matter. Nucl. Instrum. Meth. Phy. Res. B **142**, 441–462 (1998)
118. S. Datz, Atomic collision processes in solids and gases at ultrarelativistic energies. Nucl. Instr. Meth. B **164–165**, 1–11 (2000)
119. ICRU (International Commission on Radiation Units, 1993). Stopping powers and ranges for protons and alpha particles. Report 49. http://www.icru.org/
120. https://www-nds.iaea.org/stopping/
121. J. Lindhard, A.H. Sørensen, Relativistic theory of stopping for heavy ions. Phys. Rev. A **53**, 2443–2456 (1996)
122. F. Bloch, Zur bremsung rasch bewegter teilchen beim durchgang durch materie. Ann. Physik **5**, 285–320 (1933)
123. H. Walter, J. Barkas, N. Dyer, H.H. Heckman, Resolution of the S-mass anomaly. Phys. Rev. Lett. **11**, 26–28 (1963)
124. E. Fermi, The ionization loss of energy in gases and in condensed materials. Phys. Rev. **57**, 485–493 (1940)
125. C. Scheidenberger, H. Geissel, H.H. Mikkelsen, F. Nickel, T. Brohm, H. Folger, H. Irnich, A. Magel, M.F. Mohar, G. Münzenberg, M. Pfetzner, E. Roeckl, I. Schall, D. Schardt, K.-H. Schmidt, W. Schwab, M. Steiner, K. Th Stöhlker, D.J. Semerer, B. Vieira, M. Weber Voss, Direct observation of systematic deviations from the Bethe stopping theory for relativistic heavy ions. Phys. Rev. Lett. **73**, 50–53 (1994)
126. H. Weick, H. Geissel, C. Scheidenberger, F. Attallah, T. Baumann, D. Cortina, M. Hausmann, B. Lommel, G. Münzenberg, N. Nankov, F. Nickel, T. Radon, H. Schatz, K. Schmidt, J. Stadlmann, K. Smmerer, M. Winkler, H. Wollnik, Slowing down of relativistic few-electron heavy ions. Nucl. Instr. Meth. B **164–165**, 168–179 (2000)
127. J.F. Ziegler, J.B. Biersack, Computer programme SRIM, http://www.srim.org; see also: Ziegler, J.F., Ziegler, M.D., Biersack, J.P.: SRIM: The stopping and range of ions in matter. Nucl. Instrum. Methods Phy. Res. B **268**, 1818–1823 (2010)
128. Paul, H.: Stopping Power for Light Ions. http://exphys.uni-linz.ac.at/stopping
129. H. Paul, A. Schinner, An empirical approach to the stopping power of solids and gases for ions from $_3$Li to $_{18}$Ar Part II. Nucl. Ins. Meth. Phy. Res. B **195**, 166–174 (2002)
130. T.E. Pierce, M. Blann, Stopping Powers and Ranges of 5–90-MeV S, Cl, Br, and I ions in H_2, He, N_2, Ar, and Kr: A Semiempirical Stopping Power Theory for Heavy Ions in Gases and Solids. Phys. Rev. **173**, 390–405 (1968)
131. S. Nilsson, Book Review: Barkas, W.H.: Nuclear research emulsions, vol. 1, (Academic Press, New York and London, 1963), p. 178
132. J.M. Anthony, W.A. Lanford, Stopping power and effective charge of heavy ions in solids. Phys. Rev. A **25**, 1868–1879 (1982)
133. W. Brandt, M. Kitagawa, Effective stopping-power charges of swift ions in condensed matter. Phys. Rev. B **25**, 5631–5637 (1982)

134. F. Hubert, R. Bimbot, H. Gauvin, Range and stopping powertables for 2.5–500 MeV/nucleon heavy ions in solids. Atom. Data Nucl. Data Tables **46**, 1–213 (1990)
135. D.H.H. Hoffmann, J. Jacoby, W. Laux, M. de Magistris, E. Boggasch, P. Spiller, C. Stöckl, A. Tauschwitz, K. Weyrich, M. Chabot, D. Gardés, Energy loss of fast heavy ions in plasmas. Nucl. Instr. Meth. B **90**, 1–9 (1994)
136. I.L. Rachno, N.V. Mokhov, S.I. Striganov, Modeling Heavy Ion Ionization Loss in the MARS15 Code. FERMILAB-Conf-05-019-AD, *MS 220* (Batavia, IL, US, 2005), pp. 1–11
137. M.D. Barriga-Carrasco, Heavy ion charge-state distribution effects on energy loss in plasmas. Phys. Rev. E **88**, 043107–10 (2013)
138. F. Hubert, R. Bimbo, H. Gauvin, Semi-empirical formulae for heavy ion stopping powers in solids in the intermediate energy range. Nucl. Instr. Meth. B **36**, 357–363 (1989)
139. J.F. Janni, Proton range-energy Tables. 1 keV-10 GeV, Energy Loss, Range, Path Length, Time-of-Flight, Straggling, Multiple Scattering, and nuclear interaction probability. Atomic Data Nucl. Data Tables **27**, 147–339 (1982)
140. J.F. Ziegler, The electronic and nuclear stopping of energetic ions. Appl. Phys. Lett. **31**, 544–546 (1977)
141. I. Bakhmetjev, A. Cherkasov, A. Golubev, A. Fertman, V. Turtikov, B. Sharkov, V. Punin, N. Jidkov et al., Thick target approach for precise measurements of total stripping range, in *Inertial Fusion Sciences and Applications 1999*, ed. by Ch. Labaune, W.J. Hogan, K.A. Tanaka (Book of abstracts, Patis, 2000), pp. 572–575
142. I. Bakhmetjev, A. Fertman, A. Kantsyrev, V. Luckjashin, B. Sharkov, V. Turtikov, A. Kunin, V. Vatulin, N. Zhidkov, E. Vasin, U. Neuner, J. Wiesser, J. Jacoby, D.H.H. Hoffman, Research into the advanced experimental methods for precision ion stopping range measurements in matter. Laser Part. beams **21**, 1–6 (2003)
143. A. Golubev, A. Kantsyrev, V. Luckjashin, A. Fertman, A. Kunin, V. Vatulin, A. Gnutov, Y. Panov, H. Iwase, E. Mustafin, D. Schardt, K. Weyrich, N. Sobolevskiy, L. Latyshev, Measurements of the energy deposition profile for ^{238}U ions with specific energy 500 and 950 MeV/u in stainless steel and copper targets. Nucl. Instrum. Meth. B **263**, 339–344 (2007)
144. R. Bimbot, Compilation, measurements and tabulation of heavy ion stopping data. Nucl. Instrum. Meth. B **69**, 1–9 (1992)
145. H. Paul, A. Schinner, Judging the reliability of stopping power tables and programs for heavy ions. Nucl. Instrum. Meth. B **209**, 252–258 (2003)
146. W.N. Lennard, H. Geissel, D.P. Jackson, D. Phillips, Electronic stopping values for low velocity ions ($9 \leq Z_1 \leq 92$) in carbon targets. Nucl. Instrum. Meth. B **13**, 127–132 (1986)
147. G. Maynard, M. Chabot, G. Gardés, Density effect and charge dependent stopping theories for heavy ions in the intermediate velocity regime. Nucl. Instrum. Meth. B **164–165**, 139–146 (2000)
148. P. Sigmund, A. Schinner, An empirical approach to the stopping power of solids and gases for ions from ^{3}Li to ^{18}Ar - Part II. Nucl. Instrum. Meth. B **195**, 166–174 (2002)
149. A.F. Lifschitz, N.R. Arista, Effective charge and the mean charge of swift ions in solids. Phys. Rev. A **69**, 012902–5 (2004)
150. R. Bimbot, D. Gardes, H. Geissel, T. Kitahara, P. Armbruster, A. Fleury, F. Hubert, Stopping power measurements for 3.5-MeV/u Kr, Xe, Pb and U ions in solids. Nucl. Instrum. Meth. **174**, 231–236 (1980)
151. H. Geissel, Y. Laichter, W.F.W. Schneider, P. Armbruster, Energy loss and energy loss straggling of fast heavy ions in matter. Nucl. Instrum. Meth. **194**, 21–29 (1982)
152. S. Datz, H.F. Krause, C.R. Vane, H. Knudsen, P. Grafström, R.H. Schuch, Effect of nuclear size on the stopping power of ultrarelativistic heavy ions. Phys. Rev. Lett. **77**, 2925–2928 (1996)
153. W.H. Bragg, R. Kleemen, On the ionization curves of radium. Phil. Mag. S6, 726–738 (1904)
154. N. Bohr, On the decrease of velocity of swiftly moving electrified particles in passing through matter. Philos. Mag. **30**, 581–612 (1915)
155. E. Nardi, Z. Zinamon, Charge state and slowing of fast ions in a plasma. Phys. Rev. Lett. **49**, 1251–1254 (1982)

156. N.E.B. Cowern, Effective charge of energetic heavy ions in gases Solids and plasmas. J. Phys. (Paris) **44**, C8-107–C8-121 (1983)

157. M.M. Basko, Stopping of fast ions in a dense plasma. J. Plasma. Phys. **10**, 689–694 (1984). [Sov. Phys. - Fiz. Plasmy **10**, 1195–1203 (1984)]

158. J. Meyer-ter-Vehn, S. Witkowski, R. Bock, D.H.H. Hoffmann, I. Hofmann, R.W. Müller, R. Arnold, P. Mulser, Accelerator and target studies for heavy ion fusion at the Gesellschaft für Schwerionenforschung. Phys. Fluids B **2**, 1313–1317 (1990)

159. G. Zwicknagel, C. Toepffer, P.-G. Reinhard, Stopping of heavy ions in plasmas at strong coupling. Phys. Rep. **309**, 117–208 (1999)

160. K. Shibata, K. Tsubuka, T. Nishimoto, J. Hasegawa, M. Ogawa, Y. Oguri, Experimental study on the feasibility of hot plasmas as stripping media for MeV heavy ions. J. Appl. Phys. **91**, 4833–4839 (2002)

161. J. Hasegawa, N. Yokoya, Y. Kobayashi, M. Yoshida, M. Kojima, T. Sasaki, H. Fukuda, M. Ogawa, Y. Oguri, T. Murakami, Stopping power of dense helium plasma for fast heavy ions. Laser Part. Beams **21**, 7–11 (2003)

162. D.H.H. Hoffmann, K. Weyrich, H. Wahl, T. Peter, J. Meyer-terVehn, J. Jacoby, R. Bimbot, D. Gardes, M. Dumail, C. Fleurier, A. Sanba, C. Deutsch, G. Maynard, R. Noll, R. Haas, R. Arnold, S. Maurmann, Experimental observation of enhanced stopping of heavy ions in a hydrogen plasma. Z. Phys. A-Atomic Nuclei **30**, 339–345 (1988)

163. K.-G. Dietrich, D.H.H. Hoffmann, E. Boggasch, J. Jacoby, H. Wahl, M. Elfers, C.R. Haas, V.P. Dubenkov, A.A. Golubev, Charge state of fast heavy ions in a hydrogen plasma. Phys. Rev. Lett. **69**, 3623–3626 (1993)

164. G. Belyaev, M. Basko, A. Cherkasov, A. Golubev, A. Fertman, I. Roudskoy, S. Savin, B. Sharkov, V. Turtikov, A. Arzumanov, A. Borisenko, I. Gorlachev, S. Lysukhin, D.H.H. Hoffmann, A. Tauschwitz, Measurement of the Coulomb energy loss by fast protons in a plasma target. Phys. Rev. E **53**, 2701–2707 (1996)

165. M. Basko, A. Fertman, A. Golubev, M. Basko, A. Fertman, A. Kozodaev, N. Mesheryakov, B. Sharkov, A. Vishnevskiy, V. Fortov, M. Kulish, V. Gryaznov, V. Mintsev, A. Golubev, A. Pukhov, V. Smirnov, U. Funk, S. Stoewe, M. Stetter, H.-P. Flierl, D.H.H. Hoffmann, J. Jacoby, I. Iosilevski, Dense plasma diagnostics by fast proton beams. Phys. Rev. E **57**, 3363–3367 (1998)

166. V. Mintsev, V. Gryaznov, M. Kulish, A. Filimonov, V. Fortov, B. Sharkov, A. Golubev, A. Fertman, V. Turtikov, A. Vishnevskiy, A. Kozodaev, D.H.H. Hoffmann, U. Funk, S. Stoewe, M. Geissel, J. Jacoby, D. Gardés, M. Chabot, Stopping power of proton beam in a weakly non-ideal xenon plasma. Contrib. Plasma Phys. **37**, 101–104 (1999)

167. M. Roth, H. Wahl, B. Sharkov, U. Funk, M. Geissel, D.H.H. Hoffmann, A. Tauschwitz, J. Jacoby, W. Susz, U. Neuner, A. Golubev, A. Fertman, V. Turtikov, I. Roudskoy, Experimental investigation of the effective charge state of ions in beam-plasma interaction. Nucl. Instrum. Meth. A **464**, 247–252 (2001)

168. H. Wahl, M. Geissel, M. Roth, I. Roudskoy, A. Tauschwitz, I.Y. Tolstikhina, V. Turtikov, A. Fertman, D.H.H. Hoffmann, B. Sharkov, V.P. Shevelko, Interaction of fast highly charged uranium ions with a plasma target. Bull. Lebedev Phys. Inst. **8**, 28–41 (2001). (Moscow)

169. A. Fertman, T.Yu. Mutin, M.M. Basko, A.A. Golubev, T.V. Kulevoy, R.P. Kuybeda, V.I. Pershin, I.V. Roudskoy, B.Yu. Sharkov, Stopping power measurements for 100-keV/u Cu ions in hydrogen and nitrogen. Nuclear Inst. and Methods in Physics Research B **247**, 199–204 (2006)

170. M. Engelbrecht, M.P. Engelhardt, H.-P. Flierl, J. Jacoby, J. Kolb, R. Kowalewicz, A. Meineke, J. Philipps, D.H.H. Hoffmann, Numerical description and development of plasma stripper targets for heavy-ion beams. Nucl. Instrrum. Meth. A **415**, 621–627 (1998)

171. D. Gardés, A. Servajean, B. Kubica, C. Fleurier, D. Hong, C. Deutsch, G. Maynard, Stopping of multicharged ions in dense and fully ionized hydrogen. Phys. Rev. A **46**, 5101–5111 (1992)

172. S. Eisenbarth, O. Rosmej, V.P. Shevelko, A. Blazevich, D.H.H. Hoffmann, Numerical simulations of the projectile ion charge difference in sold and gaseous stopping matter. Laser and Particles Beams **27**, 601–611 (2007)

173. A. Frank, A. Blazevic, P.L. Grande, K. Harres, K. Hessuling, D.H.H. Hoffmann, R. Knobloch-Maas, P.G. Kuznetsov, F. Nürnberg, A. Pelka, G. Schaumann, G. Schiwietz, A. Schökel, M. Schollmeier, D. Schumacher, J. Schötrumpf, V.V. Vatulin, O.A. Vinokurov, M. Roth, Energy loss of argon in a laser-generated carbon plasma. Phys. Rev. E **81**, 026403–9 (2010)
174. A. Frank, A. Blazevic, V. Bagnoud, M.M. Basko, M. Börner, W. Cayzac, D. Kraus, T. Heling, D.H.H. Hoffmann, A. Ortner, A. Otten, A. Pelka, D. Pepler, D. Schumacher, An Tauschwitz, M. Roth, Energy loss and charge transfer of argon in a laser-generated carbon plasma. Phys. Rev. Lett. **110**, 115001–5 (2013)
175. N.O. Lassen, Danske Vidensk. Selsk. Mat.-Fyz. Medd. 26, No 5 (1951); ibid. 26, No 12 (1951)
176. H. Geissel, Y. Laichter, W.F.W. Schneider, P. Armbruster, On the effective charges from stopping powers of 0.5-10 MeV/u heavy ions. Phys. Lett. A **99**, 77–80 (1983)
177. N. Bohr, Scattering and stopping of fission fragments. Phys. Rev. **58**, 654655 (1940)
178. N. Bohr, Velocity-range relation for fission fragments. Phys. Rev. **59**, 270–275 (1941)
179. N. Bohr, The penetration of atomic particles through matter. Mat Fys Medd Dan Vid Selsk **18**(8), 1144 (1948)
180. N. Bohr, J. Lindhard, Electron capture and loss by heavy ions penetrating through matter. Mat Fys Medd Dan Vid Selsk **28**(7), 131 (1954)
181. H.D. Betz, L. Grodzins, Charge states and excitation of fast heavy ions passing through solids: a new model for a density effect. Phys. Rev. Lett. **25**, 211–214 (1970)
182. V.P. Shevelko, H. Tawara, O.V. Ivanov, T. Miyoshi, K. Noda, Y. Sato, A.V. Subbotin, I.Y. Tolstikhina, Target density effects in collisions of fast ions with solid targets. J. Phys. B: At. Mol. Opt. Phys. **38**, 26752690 (2005)
183. A. Fettouhi, H. Weick, M. Portillo, F. Becker, D. Boutin, H. Geissel, R.K. Knöbel, J. Kurcewicz, W. Kurcewicz, J. Kurpeta, Y. Litvinov, R.J. Livesay, D.J. Morrissey, G. Münzenberg, J.A. Nolen, H. Ogawa, N. Sakamoto, C. Scheidenberger, J. Stadlmann, M. Winkler, N. Yao, Gassolid effect in mean charge and slowing down of uranium ions at 60.2 and 200 MeV/u. Nucl. Instrum. Meth. B **245**, 32–35 (2006)
184. H. Ogawa, H. Geissel, A. Fettouhi, S. Fritzsche, M. Portillo, C. Scheidenberger, V.P. Shevelko, A. Surzhykov, H. Weick, F. Becker, D. Boutin, B. Kindler, R.K. Knöbel, J. Kurcewicz, W. Kurcewicz, B. YuA Litvinov, G. Lommel, W.R. Münzenberg, N. Plass, J. Sakamoto, H. Stadlmann, M. Tsuchida, N. Yao Winkler, Gas-solid difference in charge-changing cross sections for bare and H-like nickel ions at 200 MeV/u. Phys. Rev. A **75**, 020703R–4 (2007)
185. T. Miyoshi, K. Noda, H. Tawara, I.Y. Tolstikhina, V.P. Shevelko, Distribution of exit silicon ions over excited states after penetrating through carbon foils at 2.65, 4.3 and 6.0 MeV/u. Nucl. Instrum. Meth. B **258**, 329–339 (2007)
186. S.K. Allison, Experimental results on charge-changing collisions of hydrogen and helium at linetic energies above 0.2 keV. Rev. Mod. Phys. **30**, 1137–1168 (1958)
187. G.D. Alton, R.A. Sparrow, R.E. Olson, Plasma as a high-charge-state projectile stripping medium. Phys. Rev. A **45**, 5957–5963 (1992)
188. A. Ortner, A. Frank, A. Blazevic, M. Roth, Role of charge transfer in heavy-ion-beam-plasma interactions at intermediate energies. Phys. Rev E **91**, 023104–5 (2015)
189. S. Datz, H.O. Lutz, L.B. Bridwell, C.D. Moak, H.D. Betz, L.D. Ellsworth, Electron capture and loss cross sections of fast bromine ions in gases. PRA **2**, 430–438 (1970)
190. M. Imai, M. Sataka, K. Kawatsura, K. Takahiro, K. Komaki, H. Shibata, H. Sugai, K. Nishio, Equilibrium and non-equilibrium charge-state distributions of 2 MeV/u sulfur ions passing through carbon foils. Nucl. Instrum. Methods Phy. Res. B **267**, 26752679 (2009)
191. M. Imai, M. Sataka, M. Matsuda, S. Okayasu, K. Kawatsura, K. Takahiro, K. Komaki, H. Shibata, K. Nishio, Equilibrium and non-equilibrium charge-state distributions of 2.0 MeV/u carbon ions passing through carbon foils. Nucl. Instrum. Methods Phy. Res. B **354**, 172–176 (2015)
192. Y.A. Belkova, Y.A. Teplova, Three-component approximation for estimation of the nonequilibrium charge fractions of boron and nitrogen ions on passing through thin films. J. Surf. Invest. X-ray, Synchrotron Neutron Tech. **7**, 234–238 (2013)

193. P. Gastis, G. Perdikakis, D. Robertson, R. Almus, T. Anderson, D. Bauder, P. Collon, W. Lu, K. Ostdiek, M. Skulski, Measurement of the equilibrium charge state distributions of Ni Co, and Cu beams in Mo at 2 MeV/u: review and evaluation of the relevant semiempirical models. Nucl. Instrum. Methods Phy. Res. B **373**, 117–125 (2016)

194. W.E. Lamb Jr., Passage of uranium fission fragments through matter. Phys. Rev. **58**, 696–703 (1940)

195. K. Shima, T. Mikumo, H. Tawara, Equilibrium charge state distributions of ions ($Z_1 \geq 4$) after passage through foils: compilation of data after 1972. At. Data Nucl. Data Tables **34**, 357–391 (1986)

196. J. Khuyagbaatar, V.P. Shevelko, N. Borschevsky, C.E. Düllmann, I.I. Tolstikhina, A. Yakushev, Average charge states of heavy and superheavy ions passing through a rarified gas: theory and experiment. Phys. Rev. A **88**, 042703–8 (2013)

197. V.P. Shevelko, N. Winckler, M.S. Litsarev, Influence of multi-electron charge-changing processes on the average charge states of heavy ions passing through a He-gas target. Nucl. Instrum. Methods Phy. Res. B **330**, 82–85 (2014)

198. H.D. Betz, G. Hortig, E. Leischner, Ch. Schmelzer, B. Stadler, J. Weihrauch, The average charge of stripped heavy ions. Phys. Lett. **22**, 643–644 (1966)

199. V.S. Nikolaev, I.S. Dmitriev, On the equilibrium charge distribution in heavy element ion beams. Phys. Lett. A **28**, 277–278 (1968)

200. K. Shima, N. Kuno, M. Yamanouchi, Systematics of equilibrium charge distributions of ions passing through a carbon foil over the ranges Z = 492 and E = 0.026 MeV/u. Phys. Rev. A **40**, 3557–3570 (1989)

201. E. Baron, M. Bajard, Ch. Ricaud, Charge exchange of very heavy ions in carbon foils and in the residual gas of the GANIL cyclotrons. (In Electrostatic Accelerators and Associated Boosters, June 1992.) GANIL, BP 5027–14021 Caen-Cedex, France (1992)

202. G. Schiwietz, P.L. Grande, Improved charge-state formulas. Nucl. Instr. Meth. B **175**, 125–131 (2001)

203. W. Liu, G. Imbriani, L. Buchmann, A.A. Chen, J.M. D'Auria, A. Dónofrio, S. Engel, L. Gialanella, J. Greife, D. Hunter, A. Hussein, D.A. Hutcheon, A. Olin, D. Ottewell, D. Rogalla, J. Rogers, M. Romano, G. Roy, F. Terrasi, Charge state studies of low energy heavy ions passing through hydrogen and helium gas. Nucl. Instrum. Methods Phy. Res. A **496**, 198–214 (2003)

204. K.X. To, R. Drouin, Equilibrium charge fractions of 1 MeV/u Fe and Ni ions after passage through a carbon foil. Determination semi-empirique des etats de charges dún faisceau díons rapides ($Z \geq 18$). Nucl. Instrum. Meth. **160**, 461–463 (1979)

205. N. Miyoshi, *Penetration of Fast Positive Ions Through Carbon Foils: Analysis of Density Effects* (Ph.D, NIRS, Chiba, Japan, 2009)

206. G. Konac, Ch. Klatt, S. Kalbitzer, Universal fit formula for electronic stopping of all ions in carbon and silicon. Nucl. Instr. Meth. B **146**, 106–113 (1998)

207. J.P. Rozet, C. Stéphan, D. Vernhet, ETACHA: a program for calculating charge states at GANIL energies. Nucl. Instrum. Methods Phy. Res. B **107**, 67–70 (1996)

208. Y. Sato, A. Kitagawa, M. Muramatsu, T. Murakami, S. Yamada, C. Kobayashi, Y. Kageyama, T. Miyoshi, H. Ogawa, H. Nakabushi, T. Fujimoto, T. Miyata, Y. Sano, Charge fraction of 6.0 MeV/n heavy ions with a carbon foil: dependence on the foil thickness and projectile atomic number. Nucl. Instrum. Methods Phy. Res. B **201**, 571–580 (2003)

209. Y. Sato, T. Miyoshi, T. Murakami, K. Noda, V.P. Shevelko, H. Tawara, Penetration of 4.3 and 6.0 MeV/u highly charged, heavy ions through carbon foils. Nucl. Instrum. Methods in Phy. Res. B **225**, 439–448 (2004)

210. https://web-docs.gsi.de/~weick/charge_states/

211. http://lise.nscl.msu.edu/lise.html

212. E. Lamour, P.D. Fainstein, M. Galassi, C. Prigent, C.A. Ramirez, R.D. Rivarola, J.P. Rozet, M. Trassinelli, D. Vernhet, Extension of charge-state-distribution calculations for ion-solid collisions towards low velocities and many-electron ions. Phys. Rev. A **92**, 042703–11 (2015)

213. A. Leon, S. Melki, D. Lisfi, J.P. Grandin, P. Jardin, M.G. Suraud, A. Cassimi, Charge state distributions of swift heavy ions behind various solid targets. ($36 < Z_p < 92$, 18 MeV/u $< E < 44$ MeV/u). At. Data Nucl. Data Tables **69**, 217–238 (1998)

214. N. Winckler, A. Rybalchenko, V.P. Shevelko, M. Al-Turany, T. Kollegger, Th Stöhlker, BREIT code: analytical solution of the balance rate equations for charge-state evolutions of heavy-ion beams in matter. Nucl. Instrum. Methods Phy. Res. B **392**, 67–73 (2017)

215. http://breit.gsi.de

216. https://github.com/FAIR-BREIT/BREIT-DOC/blob/master/README.md

217. https://github.com/FAIR-BREIT/BREIT-CORE/tree/master/data/input

218. A. Frank, A. Blazevic, V. Bagnoud, M.M. Basko, M. Börner, W. Cayzac, D. Kraus, D. Herling, D.H.H. Hoffmann, A. Ortner, A. Otten, A. Pelka, D. Pepler, D. Schumacher, An Tauschwitz, M. Roth, Energy loss and charge transfer of argon in a laser-generated carbon plasma. Phys. Rev. Lett. **110**, 115001–5 (2013)

219. O. Osmani, P. Sigmund, Charge evolution of swift-heavy-ion beams explored by matrix method. Nucl. Instrum. Methods Phy. Res. B **269**, 813816 (2011)

220. Y. Zylberberg, D. Hutcheon, L. Buchmann, J. Caggiano, W.R. Hannes, A. Hussein, E. O'Connor, D. Ottewell, J. Pearson, C. Ruiz, G. Ruprecht, M. Trinczek, C. Vockenhuber, Charge-state distributions after radiative capture of helium nuclei by a carbon beam. Nucl. Instrum. Methods Phy. Res. B **254**, 1724 (2007)

221. R.C. Isler, An overview of charge-exchange spectroscopy as a plasma diagnostic. Plasma Phys. Control. Fusion **36**, 171–208 (1994)

222. C.D. Cantrell, M.O. Scully, Progress in Lasers and Laser Fusion (Studies in the Natural Sciences, vol. **8**, ed by A. Perlmutter, S.M. Widmayer, (Plenum, New York, 1975), p. 147

223. H. Daido, Review of soft x-ray laser researches and developments. Rep. Progr. Phys. **65**, 1513–1576 (2002)

224. C. Omet, P. Spiller, J. Stadlmann, D.H.H. Hoffmann, Charge change-induced beam losses under dynamic vacuum conditions in ring accelerators. New J. Phys. **8**(11), 284–302 (2006)

225. V.P. Shevelko, One-electron capture in collisions of fast ions with atoms. Z. Physik A **287**, 19–26 (1978)

226. L.P. Presnyakov, D.B. Uskov, R.K. Janev, Charge exchange in slow collisions of nulticharged ions with atoms. Sov. Phys. JETP **56**, 525–536 (1982). [Zh. Eksp. Teor. Fiz. **83**, 933-945 (1982)]

227. Y. Wu, P.C. Stancil, H.P. Liebermann, P. Funke, S.N. Rai, R.J. Buenker, D.R. Schultz, Y. Hui, I.N. Draganic, C.C. Havener, Theoretical investigation of charge transfer between N and atomic hydrogen. Phys. Rev. A **84**, 022711–8 (2011)

228. V.K. Nikulin, N.A. Guschina, Single-electron charge transfer and excitations at collisions between Bi4+ ions in the kiloelectronvolt energy range. Tech. Phys. **52**, 148–158 (2007). [Zh. Tekh. Fiz. **77**, 8–17 (2007)]

229. M.I. Chibisov, Charge exchange and ionization in the collision of atoms and multiply charged ions. JETP Lett. **24**, 46–49 (1976). [Pis'ma Zh. Eksp. Teor. Fiz. **24**, 56-60 (1976)]

230. R.E. Olson, A. Salop, Electron transfer between multicharged ions and neutral species. Phys. Rev. A **14**, 579–585 (1976)

231. A. Bárány, C.J. Setterlind, Interaction of slow highly charged ions with atoms, clusters and solids: a unified classical barrier approach. Nucl. Instrum. Meth. Phys. Res. B **98**, 184–186 (1995)

232. H. Ryufuku, T. Watanabe, Total and partial cross sections for charge transfer in collisions of multicharged ions with atomic hydrogen. Phys. Rev. A bf **20**, 1828–1837 (1979)

233. Dz. Belkic, R. Gayet, A. Salin, Computation of total cross-sections for electron capture in high energy ion-atom collisions. Comput. Phys. Commun. **23**, 153–167 (1981); ibid. **32**, 385–397 (1984)

234. I.I. Tupitsyn, Y.S. Kozhedub, V.M. Shabaev, S.B. Deyneka, S. Hagmann, C. Kozhuharov, G. Plunien, T. Stöhlker, Relativistic calculations of the charge-transfer probabilities and cross sections for low-energy collisions of H-like ions with bare nuclei. Phys. Rev. A **82**, 042701–16 (2010)

235. V.P. Shevelko, T. Stöhlker, H. Tawara, IYu. Tolstikhina, G. Weber, Electron capture in intermediate-to-fast heavy ion collisions with neutral atoms. NIMB **268**, 2611–2616 (2010)
236. L.P. Presnyakov, A.D. Ulantsev, Charge exchange between multiply charged ions and atoms Sov. J. Quantum Electron. **4**, 1320–1324 (1974). [Kvantovaya Elektron. **1**, 2377-2385 (1974)]
237. N.V. Fedorenko, Sov. Phys. - JTP **15**, 1947 (1971)
238. J.R. Macdonald, C.L. Cocke, W.W. Edison, Capture of Argon K-Shell Electrons by 2.5- to 12-MeV Protons. Phys. Rev. Lett **32**, 648–651 (1974)
239. M. Rodbro, E. Horsdal-Pedersen, J.R. Macdonald, Inner-shell-electron capture by H^+, He^{2+}, and Li^{3+} projectiles from CH_4., Ne, and Ar. Phys. Rev. A **19**, 1936–1947 (1979)
240. R.E. Olson, J. Ullrich, H. Schmidt-Böcking, Multiple-ionization collision dynamics. Phys. Rev. A **39**, 5572–5583 (1989)
241. T. Stöhlker, C. Kozhuharov, P. Mokler, R.E. Olson, Z. Stachura, A. Warczak, Single and double electron capture in collisions of highly ionized, decelerated Ge ions with Ne. J. Phys. B At. Mol. Opt. Phys. **25**, 4527–4532 (1992)
242. J.A. Perez, R.E. Olson, P.J. Beiersdorfer, Charge transfer and x-ray emission reactions involving highly charged ions and neutral hydrogen. Phys. B At. Mol. Opt. Phys. **34**, 3063–3072 (2001)
243. I. Blank, S. Otranto, C. Meinema, R.E. Olson, R. Hoekstra, State-selective electron transfer and ionization in collisions of highly charged ions with ground-state Na(3s) and laser-excited Na*(3p). Phys. Rev. A **85**, 022712–8 (2012)
244. J. Eichler, F.T. Chan, Approach to electron capture into arbitrary principal shells of energetic projectiles Phys. Rev. A **20**, 104–112 (1979)
245. V.P. Shevelko, O.N. Rosmej, H. Tawara, I.Y. Tolstikhina, The target-density effect in electron-capture processes. J. Phys. B: At. Mol. Opt. Phys. **37**, 201–213 (2004)
246. H.C. Brinkman, H.A. Kramers, Zur Theorie der Einfangung von Elektronen durch α-Teilchen. Proc. R. Acad. Sci. Amsterdam **33**, 973–984 (1930)
247. A.S. Schlachter, J.W. Stearns, W.G. Graham, K.H. Berkner, R.V. Pyle, J.A. Tanis, Electron capture for fast highly charged ions in gas targets: an impirical scaling rule. Phys. Rev. A **27**, 3372–3374 (1983)
248. H. Knudsen, H.K. Haugen, P. Hvelplund, Single-electron-capure cross section for medium- and high-velocity, highly charged ions colliding with atoms. Phys. Rev. A **23**, 597–610 (1981)
249. A. Müller, E. Salzborn, Scaling of cross sections for multiple electron transfer to highly charged ions colliding with atoms and molecules. Phys. Lett. A **62**, 391–394 (1977)
250. M. Imai, Y. Iriki, A. Itoh, Target dependence of single-electron-capture cross sections for slow Be, B, C, Fe, Ni, and W ions colliding with atomic and molecular targets. Fusion Sci. Technol. **63**, 392–399 (2013)
251. H. Knudsen, H.K. Haugen, P. Hvelplund, Single-electron capture by highly charged ions colliding with atomic and molecilar hydrogen. Phys. Rev. A **24**, 2287–2290 (1981)
252. Y. Nakai, T. Shirai, A semiempirical formula for single-electron-capture cross sections of multiply charged ions colliding with H, H_2 and He. Phys. Scripta **T28**, 77–80 (1989)
253. C. Bottcher, *Coherence and correlation in atomic collisions*, ed. by H. Kleinpoppen, J.F. Williams Plenum, (New York, 1980), pp. 403–409
254. V.P. Shevelko, P. Scharrer, Ch. E. Düllmann, W. Barth, J. Khuyagbaatar, I.Yu. Tolstikhina, N. Winckler, A. Yakushev, Charge-state dynamics of 1.4 and 11 MeV/u uranium ions penetrating H_2 and He gas targets, Nucl. Instrum. Methods B (2018, to be submitted)
255. V.P. Shevelko, N. Winckler, IYu. Tolstikhina, Gas-pressure dependence of charge-state fractions and mean charges of 1.4 MeV/u-uranium ions stripped in molecular hydrogen. Nucl. Instrum. Methods Phy. Res. B **377**, 77–82 (2016)
256. H. Klinger, A. Müller, E. Salzborn, Electron capture processes of multiply charged argon ions in argon at energies from 10 to 90 keV. J. Phys. B At. Mol. Phys. **8**, 230–238 (1975)
257. D.H. Crandall, R.E. Olson, E.J. Shipsey, J.C. Browne, Single and double charge transfer in C^{4+}-He collisions. Phys. Rev. Lett. **36**, 858–860 (1976)
258. D.H. Crandall, R.A. Phaneuf, F.W. Meyer, Electron capture by heavy multicharged ions from atomic hydrogen at low velocities. Phys. Rev. A **22**, 379–387 (1980)

259. H. Tawara, T. Iwai, Y. Kaneko, M. Kimura, N. Kobayashi, A. Matsumoto, S. Ohtani, K. Okuno, S. Takagi, S. Tsurubuchi, Electron capture processes of I^{q+} ions with very high charge states in collisions with He atoms. J. Phys. B At. Mol. Phys. **18**, 337–350 (1985)

260. A. Müller, R. Schuch, W. Groh, E. Salzborn, H.F. Beyer, P.H. Mokler, R.E. Olson, Multiple-electron capture and ionization in collisions of highly stripped ions with Ar atoms. Phys. Rev. A **33**, 3010–3017 (1986)

261. N. Selberg, C. Biedermann, H. Cederquist, Semiempirical scaling laws for electron capture at low energies. Phys. Rev. A **54**, 4127–4135 (1996)

262. N. Selberg, C. Biedermann, H. Cederquist, Absolute charge-exchange cross sections for the interaction between slow Xe^{q+} ($15 \leq q \leq 43$) projectiles and neutral He, Ar, and Xe. Phys. Rev. A **56**, 4623–4632 (1997)

263. M. Kimura, N. Nakamura, H. Watanabe, I. Yamada, A. Danjo, K. Hosaka, A. Matsumoto, S. Ohtani, H.A. Sakaue, M. Sakurai, H. Tawara, M. Yoshino, A scaling law of cross sections for multiple electron transfer in slow collisions between highly charged ions and atoms. J. Phys. B At. Mol. Opt. Phys. **28**, L643–L647 (1995)

264. N. Nakamura, F.J. Currell, A. Danjot, M. Kimuras, A. Matsumotoll, S. Ohtani, H.A. Sakaue, M. Sakurai, H. Tawara, H. Watanabet, I. Yamada, M. Yoshino, Target dependence of multi-electron processes in I^{q+} ($q = 10, 15$) +rare gas (Ne, Ar, Kr and Xe) collisions. J. Phys. B At. Mol. Opt. Phys. **28**, 2959–2972 (1995)

265. K. Suzuki, K. Okuno, N. Kobayashi, Single- and multiple-charge changing cross sections in slow collisions of Ar^{q+} ($q = 4$–9) with Ne. Phys. Scripta **T73**, 172–174 (1997)

266. J.H. McGuire, *Introduction to Dynamic Correlation: Multiple Electron Transitions in Atomic Collisions* (Tulane University, New Orleans, 1997)

267. H.A. Sakaue, A. Danjo, K. Hosaka, D. Kato, M. Kimura, A. Matsumoto, N. Nakamura, S. Ohtani, M. Sakurai, H. Tawara, I. Yamada, M. Yoshino, Electron transfer and decay processes of highly charged iodine ions. J. Phys. B At. Mol. Opt. Phys. **37**, 403–415 (2004)

268. H. Brauning, R. Trassl, A. Diehl, A. Theiss, E. Salzborn, A.A. Narits, L.P. Presnyakov, Resonant electron transfer in collisions between two fullerene ions. Phys. Rev. Lett. **91**, 168301–4 (2003)

269. H. Zettergren, P. Reinhed, K. Stochkel, H.T. Schmidt, H. Cederquist, J. Jensen, S. Tomita, B. Nielsen, P. Hvelplund, B. Manil, J. Rangama, B.A. Huber, Fragmentation and ionization of C_{70} and C_{60} by slow ions of intermediate charge. Eur. Phys. J. D **38**, 299–306 (2006)

270. S. Wethekam, H. Zettergren, Ch. Linsmeier, H. Cederquist, H. Winter, Resonant electron capture by C60 ions at a metal surface with projected band gap. Phys. Rev. B **81**, 121416–4(R) (2010)

271. A.A. Narits, Charge transfer between fullerenes and highly charged noble gas ions. J. Phys. B At. Mol. Opt. Phys. **41**, 135102–10 (2008)

272. A. Bárány, G. Astner, H. Cederquist, H. Danared, H. Huldt, P. Hvelplund, A. Johnson, H. Knudsen, L. Liljeby, K.-G. Rensfelt, Absolute cross sections for multi-electron processes in low energy Ar^{q+}-Ar collisions: comparison with thoery. Nucl. Instrum. Meth. Phys. Res. B **9**, 397–399 (1985)

273. A.J. Niehaus, A classical model for multiple-electron capture in slow collisions of highly charged ions with atoms. Phys. B At. Mol. Phys. **19**, 2925–2937 (1986)

274. C. Harel, H. Jouin, Double capture into autoionizing states in I^{q+}-He collisions at low impact energies. J. Phys. B At. Mol. Opt. Phys. **25**, 221–237 (1992)

275. Z. Chen, R. Shingal, C.D. Lin, State-selective double capture in collisions of bare ions with helium atoms at low energies. I. Total cross sections. J. Phys. B At. Mol. Opt. Phys. **24**, 4215–4230 (1991)

276. J.P. Hansen, K. Taulbjerg, Electron correlation in highly-charged-ion collisions. Phys. Rev. A **45**, R4214–R4217 (1992)

277. I. Dź Belkić, Mančev, J. Hanssen, Four-body methods for high-energy ion-atom collisions. Rev. Mod. Phys. **80**, 249–314 (2008)

278. V.K. Nikulin, N.A. Guschina, Theory of double electron capture in slow collisions involving multiply charged ions. Phys. Scripta **T71**, 134–139 (1997)

279. V.A. Sidorovich, V.S. Nikolaev, J.H. McGuire, Calculation of the charge-changing cross sections in collisions of H^+, He^{2+} and Li^{3+} with He atoms. Phys. Rev. A **31**, 2193–2201 (1985)
280. I. Dź, Belkić, Mančev, M. Mudrinic, Two-electron capture from helium by fast α particles. Phys. Rev. A **49**, 3646–3658 (1994)
281. IYu. Tolstikhina, O.I. Tolstikhin, H. Tawara, Shake-off mechanism of two-electron transitions in slow ion-atom collisions. Phys. Rev. A **57**, 4387–4393 (1998)
282. C.R. Vane, S. Datz, H.F. Krause, P.F. Dittner, E.F. Deveney, H. Knudsen, Per Grafström, R. Schuch, H. Gao, R. Hutton, Atomic collisions with 33-TeV lead ions. Physica Scripta **T173**, 167–171 (1997)
283. R. Raisbeck, F. Yiou, Electron capture by 40−, 155− and 600-MeV ptorons in thin foils of Mylar, Al. Ni and Ta. Phys. Rev. A **4**, 1858–1868 (1971)
284. H.W. Schnopper, H.D. Betz, J.P. Delvaille, K. Kalata, A.R. Sohval, K.W. Jones, H.E. Wegner, Evidence for radiative electron capture by fast, highly stripped heavy ions. Phys. Rev. Lett. **29**, 898–901 (1972)
285. P. Kienle, M. Kleber, B. Povh, R.M. Diamond, F.S. Stephens, E. Grosse, M.R. Maier, D. Proetel, Radiative capture and bremsstrahlung of bound electrons induced by heavy ions. Phys. Rev. Lett. **31**, 1099–1102 (1973)
286. J.K.M. Eichler, Eikonal theory of charge exchange between arbitrary hydrogenic states of target and projectile. Phys. Rev. A **23**, 498–509 (1981)
287. J. Eichler, Relativistic eikonal theory of electron capture. Phys. Rev. **32**, 112–121 (1985)
288. A. Ichihara, T. Shirai, J. Eichker, Radiative electron capture in relativistic atomic collisions. Phys. Rev. A **49**, 1875–1884 (1994)
289. J. Eichler, W.E. Meyerhof, *Relativistic Atomic Collisions* (Academic Press, Sab Diego, 1995)
290. A. Ichihara, T. Shirai, J. Eichler, Cross sections for electron capture in relativistic atomic collisions. Atom. Data Nucl. Data Tables **55**, 63–79 (1993)
291. A. Ichihara, J. Eichler, Cross sections for radiative recombination and the photoelectric effect in the K, L, and M shells of one-electron systems with $1 \leq Z \leq 112$ calculated within an exact relativistic description. Atomic Data Nucl. Data Tables **74**, 1121 (2000)
292. M. Stobbe, Zur Quantenmechanik photoelektrischer Prozesse. Ann. Phys. **7**, 661–715 (1930)
293. J. Eichler, Theory of relativistic charge exchange with Coulomb boundary conditions. Phys. Rev. A **35**, 3248–3255 (1987)
294. J. Eichler, Erratum: Theory of relativistic charge exchange with Coulomb boundary conditions [Phys. Rev. A **35**, 3248 (1987)]. Phys. Rev. A **37**, 287 (1988)
295. T. Th Stöhlker, H. Ludziejewski, F. Reich, R.W. Bosch, J. Dunford, B. Eichler, C. Franzke, G. Kozhuharov, P.H. Menzel, F. Mokler, P. Nolden, Z. Rymuza, M. Stachura, P. Steck, A. Swiat, T. Winkler, Warczak, Charge-exchange cross sections and beam lifetimes for stored and decelerated bare uranium ions. Phys. Rev. A **58**, 2043–2050 (1998)
296. Th. Stöhlker, O. Brinzanescu, A. Krämer, T. Ludziejewski, X. Ma, A. Warczak, in *X-Ray and Inner Shell Processes; 18th International Conference, AIP Conference Proceedings 506*, ed by D.S. Gemmel, E.P. Kanter, L. Young, Chicago, Illinois, p. 389, 1999 (2000)
297. S. Fritzsche, A. Surzhykov, Th Stöhlker, Radiative recombination into high-Z few-electron ions: cross sections and angular distributions. Phys. Rev. A **72**, 012704–11 (2005)
298. T. Th Stöhlker, F. Ludziejewski, R.W. Bosch, C. Dunford, P.H. Kozhuharov, H.F. Mokler, O. Beyer, B. Brinzanescu, J. Franzke, A. Eichler, S. Griegal, A. Hagmann, A. Ichihara, D. Krämer, F. Lekki, H. Liesen, P. Nolden, Z. Reich, M. Rymuza, J. Stachura, P. Steck, A. Warczak Swiat, Angular distribution studies for the time-reversed photoionization process in hydrogenlike uranium: the identification of spin-flip transitions. Phys. Rev. Lett. **82**, 3232–3235 (1999)
299. N. Stolterfoht, R. Cabrera-Trujillo, Y. Öhrn, E. Deumens, R. Hoekstra, J.R. Sabin, Strong isotope effects on the charge transfer in slow collisions of $He^{?+}$ with atomic hydrogen, deuterium, and tritium. Phy. Rev. Lett. **99**(10), 103201–4 (2007)
300. H. Bolt, V. Barabash, W. Krauss, J. Linke, R. Neu, S. Suzuki, N. Yoshida, A.U. Team, Materials for the plasma-facing components of fusion reactors. J. Nucl. Mater. **329–333**, 66–73 (2004)

301. Y. Ralchenko, I.N. Draganic, J.N. Tan, J.D. Gillaspy, J.M. Pomeroy, J. Reader, U. Feldman, G.E. Holland, EUV spectra of highly-charged ions W^{54+}-W^{63+} relevant to ITER diagnostics. J. Phy. B: At., Mol. Opt. Phy. **41**(2), 021003–6 (2008)
302. I.Y. Tolstikhina, M.Y. Song, M. Imai, Y. Iriki, A. Itoh, D. Kato, H. Tawara, J.S. Yoon, V.P. Shevelko, Charge-changing collisions of tungsten and its ions with neutral atoms. J. Phy. B: At., Mol. Opt. Phy. **45**(14), 145201–8 (2012)
303. I.Y. Tolstikhina, O.I. Tolstikhin, Effect of electron-nuclei interaction on internuclear motions in slow ion-atom collisions. Phy. Rev. A **92**(4), 042707–5 (2015)
304. S. Kado, K. Suzuki, Y. Iida, A. Muraki, Doppler and Stark broadenings of spectral lines of highly excited helium atoms for measurement of detached recombining plasmas in MAP-II divertor simulator. J. Nucl. Mater. **415**(1), S1174–S1177 (2011)
305. S. Kado, Y. Iida, S. Kajita, D. Yamasaki, A. Okamoto, B. Xiao, T. Shikama, T. Oishi, S. Tanaka, Diagnostics of recombining plasmas in divertor simulator MAP-II. J. Plasma Fusion Res. **81**(10), 810–821 (2005)
306. I.Y. Tolstikhina, D. Kato, V.P. Shevelko, Influence of the isotope effect on the charge exchange in slow collisions of Li, Be, and C ions with H, D, and T. Phys. Rev. A **84**(1), 012706–6 (2011)
307. I. Komarov, L. Ponomarev, S. Slavyanov, *Spheroidal and Coulomb Spheroidal Functions* (Nauka, Moscow, 1976) (in Russian)
308. E.A. Solovev, *Workshop on Hidden Crossings in Ion-Collisions and in Other Nonadiabatic Transitions* (Harvard Smithonian Centre for Astrophysics, Cambridge, MA, 1991)
309. T.P. Grozdanov, E.A. Solov'ev, Charge exchange, excitation, and ionization via hidden avoided crossings. Phys. Rev. A **42**(5), 2703–2718 (1990)
310. N. Stolterfoht, R. Cabrera-Trujillo, P.S. Krstić, R. Hoekstra, Y. Öhrn, E. Deumens, J.R. Sabin, Isotope effects on the charge transfer into the n = 1, 2, and 3 shells of He^{2+} in collisions with H, D, and T. Phy. Rev. A **81**(5), 052704 (2010)
311. R. Cabrera-Trujillo, J.R. Sabin, Y. Öhrn, E. Deumens, N. Stolterfoht, Mass scaling laws due to isotopic effects in the energy loss of He^{2+} colliding with H, D, and T. Phy. Rev. A **83**(1), 012715–8 (2011)
312. E. Deumens, A. Diz, R. Longo, Y. Öhrn, Time-dependent theoretical treatments of the dynamics of electrons and nuclei in molecular systems. Rev. Mod. Phys. **66**(3), 917–983 (1994)
313. C.N. Liu, A.T. Le, T. Morishita, B.D. Esry, C.D. Lin, Hyperspherical close-coupling calculations for charge-transfer cross sections in He^{2+} + H(1s) collisions at low energies. Phy. Rev. A **67**(5), 052705–12 (2003)
314. W. Seim, A. Müller, I. Wirkner-Bott, E. Salzborn, Electron capture by Li^{i+} (i=2,3), Ni^{i+} and Ne^{i+} (i=2,3,4,5) ions from atomic hydrogen. J. Phy. B: At. Mol. Phy. **14**(18), 3475–3491 (1981)
315. D. Belkić, R. Gayet, A. Salin, Computation of total cross-sections for electron capture in high energy ion-atom collisions. Comput. Phy. Commun. **23**(2), 153–167 (1981)
316. K. Ikeda, Progress in the ITER physics basis. Nucl. Fusion **47**(6), S1–S413 (2007)
317. M. Imai, T. Shirai, M. Saito, Y. Haruyama, A. Itoh, N. Imanishi, F. Fukuzawa, H. Kubo, Production and compilation of charge changing cross sections of ion-atom and ion-molecule collisions. J. Plasma Fusion Res. Ser. **37**, 323–326 (2006)
318. M. Khoma, M. Imai, O. Karbovanets, Y. Kikuchi, M. Saito, Y. Haruyama, M. Karbovanets, I. Kretinin, A. Itoh, R. Buenker, A simple theoretical approach of charge transfer processes in collisions of atomic ions with polar targets. Chem. Phys. **352**(1–3), 142–146 (2008)
319. R.W. McCullough, W.L. Nutt, H.B. Gilbody, One-electron capture by slow doubly charged ions in H and H_2. J. Phy. B: At. Mol. Phys. **12**(24), 4159–4169 (1979)
320. F.W. Meyer, R.A. Phaneuf, H.J. Kim, P. Hvelplund, P.H. Stelson, Single-electron-capture cross sections for multiply charged O, Fe, Mo, Ta, W, and Au ions incident on H and H_2 at intermediate velocities. Phys. Rev. A **19**, 515–525 (1979)
321. H.B. Gilbody, Charge transfer and ionization in collisions with hydrogen atoms. Phy. Scr. **24**(4), 712–724 (1981)
322. M. Gargaud, R. McCarroll, M.A. Lennon, S.M. Wilson, R.W. McCullough, H.B. Gilbody, One-electron capture by slow Al^{2+} ions in atomic and molecular hydrogen. J. Phy. B: At. Mol. Opt. Phy. **23**(3), 505–511 (1990)

323. I.Y. Tolstikhina, D. Kato, Resonance charge exchange between excited states in slow proton-hydrogen collisions. Phy. Rev. A **82**(3), 032707–6 (2010)

324. L. Landau, E. Lifschitz, *Mechanics* (Butterworth-Neinemann, Oxford, 1977)

325. O.B. Firsov, Resonant charghe exchange between ions at slow collisions. JETP **21**, 1001–1008 (1951)

326. Y.N. Demkov, Charge transfer at small resonance defects. Zh. Eksp. Teor. Fiz. **45**, 195–201 (1963). [Sov. Phys. JETP **18**, 138-142 (1964)]

327. A.V. Matveenko, L.P., Orbiting-type oscillations in the total charge exchange cross section for the p + H(1s) → H(1s) + p. Zh. Eksp. Teor. Fiz. **68**(3), 920–928 (1975). [Sov. Phys.- JETP **41**, 456–459 (1976)]

328. G. Hunter, M. Kuriyan, Proton collisions with hydrogen atoms at low energies: quantum theory and integrated cross-sections. Proc. R. Soc. London A: Math. Phys. Eng. Sci. **353**(1675), 575–588 (1977)

329. P.S. Krstić, J.H. Macek, S.Y. Ovchinnikov, D.R. Schultz, Analysis of structures in the cross sections for elastic scattering and spin exchange in low-energy $H^+ + H$ collisions. Phys. Rev. A **70**(4), 042711–8 (2004)

330. S.A. Blanco, C.A. Falcon, R.D. Piacentini, Electron capture from H(2s) by H^+ at low energies. J. Phy. B: At. Mol. Phy. **19**(23), 3945–3950 (1986)

331. N. Rosen, C. Zener, Double Stern-Gerlach Experiment and Related Collision Phenomena. Phys. Rev. **40**, 502–507 (1932)

332. S. Ovchinnikov, E. Solov'ev, Theory of nonadiabatic transitions in a system of three charged particles. Zh. Eksp. Teor. Fiz. **90**, 921–925 (1986). [Sov. Phys. JETP **63**, 538-544 (1986)]

333. R.K. Janev, Hidden crossing nature of nonadiabatic coupling between quasiresonant one-electron molecular states. Phys. Rev. A **55**, R1573–R1576 (1997)

334. A. Igarashi, C.D. Lin, Full ambiguity-free quantum treatment of $D^+ + H(1s)$ charge transfer reactions at low energies. Phys. Rev. Lett. **83**, 4041–4044 (1999)

335. J.H. Macek, P.S. Krstić, S.Y. Ovchinnikov, Regge oscillations in integral cross sections for proton impact on atomic hydrogen. Phys. Rev. Lett. **93**, 183203–4 (2004)

336. V.N. Ostrovsky, Mechanisms of the rearrangement processes in the dtμ system: nonadiabatic transitions and interference effects. Phys. Rev. A **61**, 032505–9 (2000)

337. O.I. Tolstikhin, C. Namba, Hyperspherical calculations of low-energy rearrangement processes in dtμ. Phys. Rev. A **60**, 5111–5114 (1999)

338. B.D. Esry, Z. Chen, C.D. Lin, R.D. Piacentini, Close-coupling calculations of electron capture cross sections from the n = 2 states of H by protons and alpha particles. J. Physics B: At. Mol Opt Phy. **26**(9), 1579–1586 (1993)

339. Y.N. Demkov, C.V. Kunasz, V.N. Ostrovskii, United-atom approximation in the problem of $\Sigma - \Pi$ transitions during close atomic collisions. Phys. Rev. A **18**, 2097–2106 (1978)

340. V. Abramov, F. Baryshnikov, V. Lisitsa, Charge transfer between hydrogen atoms and the nuclei of multicharged ions with allowance for the degeneracy of the final states. Zh. Eksp. Teor. Fiz. **74**, 897–904 (1978). [Sov. Phys. JETP **47**, 469–477 (1979)]

341. R.K. Janev, D.S. Belić, B.H. Bransden, Total and partial cross sections for electron capture in collisions of hydrogen atoms with fully stripped ions. Phys. Rev. A **28**, 1293–1302 (1983)

342. J. Macek, X.Y. Dong, Calculation of electron-capture cross sections in low-energy collisions of C^{6+} with H. Phys. Rev. A **40**, 95–100 (1989)

343. R.L. Watson, Y. Peng, V. Horvat, G.J. Kim, R.E. Olson, Target Z-dependence and additivity of cross sections for electron loss by 6-MeV/amu Xe^{18+} projectiles. Phys. Rev. A **67**, 022706–7 (2003)

344. V.P. Shevelko, M.S. Litsarev, M.-Y. Song, H. Tawara, J.-S. Yoon, Electron loss of fast many-electron ions colliding with neutral atoms: possible scaling rules for the total cross sections. J. Phys. B: At. Mol. Opt. Phys. **42**, 065202–6 (2009)

345. J. Alonso, H. Gould, Charge-changing cross sections for Pb and Xe ions at velocities up to 4 $\times 10^9$ cm/sec. Phys. Rev. A **26**, 1134–1137 (1981)

346. H.-P. Hülskötter, B. Feinberg, W.E. Meyerhof, A. Belkacem, J.R. Alonso, L. Blumenf, E.A. Dillard, H. Gould, N. Guardala, G.F. Kreb, A. McMahan, M.F. Rhodes-Brown, B.S. Rude, J.

Schweppe, D.W. Spooner, K. Street, P. Thieberger, H.E. Wegner, Electron-electron interaction in projectile electron loss. Phys. Rev. A **44**, 1712–1724 (1991)

347. N. Madsen, Vacuum changes during accumulations of Pb^{54+} in LEIR PS/DI Note 99–21 (1999)

348. O. Gröbner, et al., GSI/CERN collaboration (1993, unpublished)

349. D. Mueller, L. Grisham, I. Kaganovich, R.L. Watson, V. Horvat, K.E. Zaharakis, Y. Peng, Multiple electron stripping of heavy ion beams. Laser Part. Beams **20**, 551–554 (2002)

350. V.P. Shevelko, IYu. Tolstikhina, Th Stöhlker, Stripping of fast heavy low-charged ions in gaseous targets. Nucl. Instrum. Methods B **184**, 295–308 (2001)

351. W.E. Meyerhof, R. Anholt, J. Eichler, H. Gould, Ch. Munge, J. Alonso, P. Thieberger, H.E. Wegner, Atomic collisions with relativistic heavy ions. III. Electron capture. Phys. Rev. A **32**, 3291–3301 (1985)

352. R. Anholt, W.E. Meyerhof, H. Gould, Ch. Munge, J. Alonso, P. Thieberger, H.E. Wegner, Atomic collisions with relativistic heavy ions. IV. Projectile K-shell ionization. Phys. Rev. A **32**, 3302–3309 (1985)

353. R. Anholt, W.E. Meyerhof, Atomic collisions with relativistic heavy ions. V. The states of ions in matter. Phys. Rev. A **33**, 1556–1568 (1986)

354. R. Anholt, Ch. Stoller, J.D. Molitoris, D.W. Spooner, E. Morenzoni, S.A. Andriamonje, W.E. Meyerhof, H. Bowman, J.-S. Xu, Z.-Z. Xu, J.O. Rasmussen, D.H.H. Hoffmann, Phys. Rev. A **33**, 2270–2280 (1986)

355. R. Anholt, U. Becker, Atomic collisions with relativistic heavy ions. IX. Ultrarelativistic collisions. Phys. Rev. A **36**, 4628–4636 (1987)

356. H. Th Stöhlker, H. Geissel, C. Folger, P.H. Kozhuharov, G. Mokler, D.Schardt Münzenberg, M. Th Schwab, H. Steiner, K.Summerer Stelzer, Equilibrium charge state distributions for relativistic heavy ions. Nucl. Instrum. Meth. B **61**, 408–410 (1991)

357. C. Scheidenberger, H. Geissel, H. Th Stöhlker, H. Folger, C. Irnich, A. Kozhuharov, P.H. Magel, R. Mokler, G. Moshammer, F. Münzenberg, M. Nickel, P. Pfätzner, W. Rymuza, J. Schwab, B.Voss Ullrich, Charge states and energy loss of relativistic heavy ions in matter. Nucl. Instr. Meth. B **90**, 36–40 (1994)

358. C. Th Stöhlker, P.H. Kozhuharov, A. Mokler, F. Warczak, H. Bosch, R. Geissel, C. Moshammer, J. Scheidenberger, A. Eichler, T. Ichihara, Z. Shirai, P. Rymuza Stachura, Radiative electron capture studied in relativistic heavy-ionatom collisions. Phys. Rev. A **51**, 2098–2111 (1995)

359. D.C. Th Stöhlker, P. Ionescu, F. Rymuza, H. Bosch, C. Geissel, T. Kozhuharov, P.H. Ludziejewski, C. Mokler, Z. Scheidenberger, A. Stachura, R.W. Dunford Warczak, K-shell excitation studied for H- and He-like bismuth ions in collisions with low-Z target atoms. Phys. Rev. A **57**, 845–854 (1998)

360. A.C.F. Santos, R.D. DuBois, Scaling laws for single and multiple electron loss from projectiles in collisions with a many-electron target. Phys. Rev. A **69**, 042709–11 (2004)

361. R.D. DuBois, A.C.F. Santos, F. Th Stöhlker, A. Bosch, A. Brauning-Demian, S. Gumberidze, C. Hagmann, R. Kozhuharov, A. Orsic Mann, U. Muthig, S. Spillmann, W. Tachenov, L. Barth, B.B. Dahl, J. Franzke, L. Glatz, R. Groning, D. Richter, K. Wilms, O. Jagutzki Ullmann, Electron loss from 1.4-MeV/u $U^{4,6,10+}$ ions colliding with Ne, N_2, and Ar targets. Phys. Rev. A **70**, 032712–5 (2004)

362. R.E. Olson, R.L. Watson, V. Horvat, K.E. Zaharakis, R.D. DuBois, Th Stöhlke, Electron stripping cross-sections for fast, low charge state uranium ions. Nucl. Instrum. Methods A **544**, 333–336 (2005)

363. R.D. DuBois, A.C.F. Santos, R.E. Olson, Scaling laws for electron loss from ion beams. Nucl. Instrum. Methods A **544**, 497–501 (2005)

364. V.I. Matveev, D.B. Sidorov, Energy losses of fast heavy multiply charged structural ions in collisions with complex atoms. JTP Techn. Phys. **52**, 839–844 (2007). [Zh. Tech. Fiz. **77**, 18–21 (2007)]

365. IYu. Tolstikhina, I.I. Tupitsyn, S.N. Andreev, V.P. Shevelko, Influence of relativistic effects on electron-loss cross sections of heavy and superheavy ions colliding with neutral atoms. JETP **119**, 1–7 (2014). [Zh. Eks. Teor. Fyz. **146**, 5-12 (2014)]

366. R.E., Olson, *Multiple Electron Capture and Ionization in Ion-Atom Collisions in Electronic and Atomic Collisions*, ed. by H.B. Gilbody, W.R. Newell, F.H. Read, A.C.H. Smith (Elsevier, New York, 1988), pp. 271–285

367. R.E. Olson, Classical trajectory and monte carlo techniques, in *Atomic, Molecular, and Optical Physics Handbook*, ed. Drake, G.W.F. AIP, Woodbury, NY, (1996), Chap. 56, pp. 869–874

368. V.P. Shevelko, D. Kato, M.S. Litsarev, H. Tawara, The energy-deposition model: electron loss of heavy ions in collisions with neutral atoms at low and intermediate energies. J. Phys. B: At. Mol. Opt. Phys. **43**, 215202–9 (2010)

369. C.L. Cocke, Production of highly charged low-velocity recoil ions by heavy-ion bombardment of rare-gas targets. Phys. Rev. A **20**, 749–758 (1979)

370. F. Salvat, J.D. Martínes, R. Mayol, J. Parellada, Analytical Dirac-Hartree-Fock-Slater screening function for atoms (Z = 1–92). Phys. Rev. A **36**, 467–474 (1987)

371. A. Russek, J. Meli, Ijnization phenomena in high-energy atomic collisions. Physica **46**, 222–243 (1970)

372. T.A. Carlson, C.W. Nestor Jr., N. Wasserman, J.P. McDowell, Calculated ionization potentials for multiply charged ions. Atom. Data Nucl. Data Tables **2**, 63–99 (1970)

373. J.P. Desclaux, Relativistic Dirac-Fock expectation values for atoms with Z = 1 to Z = 120. Atom. Data Nucl. Data Tables **12**, 311–406 (1973)

374. K. Rashid, M.Z. Saadi, M. Yasin, Dirac-Fock total energies, ionization energies, and orbital energies for uranium ions U I to U XCII. Atom. Data Nucl. Data Tables **40**, 365–378 (1988)

375. G.H. Zschornack, *Handbook of X-Ray Data* (Springer, Heidelberg, 2007)

376. N.M. Kabachnik, V.N. Kondratyev, Z. Roller-Lutz, H.O. Lutz, Multiple ionization of atoms and molecules in collisions with fast ions: ion-atom collisions. Phys. Rev. A **56**, 2848–2854 (1997)

377. V.P. Shevelko, M.S. Litsarev, H. Tawara, Multiple ionization of fast heavy ions by neutral atoms in the energy deposition model. J. Phys. B: At. Mol. Opt. Phys. **41**, 115204–5 (2008)

378. J.M. Rost, T. Pattard, Analytical parametrization for the shape of atomic ionization cross sections. Phys. Rev. A **55**, R5–R7 (1997)

379. L. Bozyk, F. Chill, M.S. Litsarev, IYu. Tolstikhina, V.P. Shevelko, Multiple-electron losses in uranium ion beams in heavy ion synchrotrons. Nucl. Instrum. Methods Phys. Res. B. **372**, 102–108 (2016)

380. V.P. Shevelko, M.-Y. Song, IYu. Tolstikhina, H. Tawara, J.-S. Yoon, Cross sections for charge-changing collisions of many-electron uranium ions with atomic and molecular targets. Nucl. Instrum. Methods Phys. Res. B. **278**, 63–69 (2012)

381. M.-Y. Song, M.S. Litsarev, V.P. Shevelko, H. Tawara, J.-S. Yoon, Single- and multiple-electron loss cross-sections for fast heavy ions colliding with neutrals: semi-classical calculations. Nucl. Instrum. Methods Phys. Res. B **267**, 2369–2375 (2009)

382. V.P. Shevelko, M.S. Litsarev, H. Th Stöhlker, Tawara, IYu. Tolstikhina, G. Weber, Electron loss and capture processes, in *collisions of heavy many-electron ions with neutral atoms, in Atomic Processes in Basic and Applied Physics*, ed. by V. Shevelko, H. Tawara (Springer, Berlin, 2012), pp. 125–152

383. A.B. Voitkiv, Theory of projectile-electron excitation and loss in relativistic collisions with atoms. Phys. Rep. **392**, 191–277 (2004)

384. A.B. Voitkiv, B. Najjari, A. Surzhykov, Charge states and effective loss cross sections for 33 TeV lead ions penetrating aluminum and gold foils. J. Phys. B **41**, 111001–7 (2008)

385. B. Najjari, A. Surzhykov, A.B. Voitkiv, Relativistic time dilation and the spectrum of electrons emitted by 33-TeV lead ions penetrating thin foils. Phys. Rev. A **77**, 042714–5 (2008)

386. A.B. Voitkiv, J. Ullrich, *Collisions of Structured Atomic Particles* (Springer, Heidelberg, 2008)

387. G. Baur, I.L. Beigman, I.Y. Tolstikhina, V.P. Shevelko, Th Stöhlker, Ionization of highly charged relativistic ions by neutral atoms and ions. Phys. Rev. A **80**, 012713–6 (2009)

388. E. Eliav, U. Kaldor, P. Schwerdtfeger, B.A. Hess, Y. Ishikawa, Ground state electron configuration of element 111. Phys. Rev. Lett. **73**, 3203–3206 (1994)

389. V.P. Shevelko, Yu.A. Litvinov, Th. Stöhlker, I.Yu. Tolstikhina, *Lifetimes of relativistic heavy-ion beams in the high energy storage ring of FAIR* (Nucl. Instrum, Methods B, 2018). (in press)

390. T.P. Grozdanov, R.K. Janev, Electron capture in slow collisions of multiply charged ions with hydrogen molecules. J. Phys. B **13**, L69–L72 (1980)
391. L. Meng, C.O. Reinhold, R.E. Olson, Electron removal from molecular hydrogen by fully stripped ions at intermediate energies. Phys. Rev. A **40**, 3637–3645 (1989)
392. D.H. Crandall, R.A. Phaneuf, F.W. Meyer, Electron capture by slow multicharged ions in atomic and molecular hydrogen. Phys. Rev. A **19**, 504–514 (1979)
393. T. Miyoshi, K. Noda, Y. Sato, H. Tawara, I.Y. Tolstikhina, V.P. Shevelko, Evaluation of excited nl-state distributions of fast exit ions after penetrating through solid foils. Part 1: charge-state fractions for 4.3 MeV/u projectiles with atomic numbers Z = 6–26 passing through carbon foils. Nucl. Instrum. Methods Phy. Res. B **251**, 79–88 (2006)
394. R.L. Hickok, Plasma density measurement by molecular ion breakup. Rev. Sci. Instrum. **38**, 142–143 (1967)
395. A. Müller, Electron-ion collisions: fundamental processes in the focus of applied research. Adv. At. Mol. Opt. Phys. **55**, 294–417 (2008)
396. I.I. Sobelman, *Atomic Spectra and Radiative Transitions* (Springer, Berlin, 1979)
397. Y.S. Kim, R.H. Pratt, Direct radiative recombimation of electrons with atomic ions: cross sections ans rate coefficients. Phys. Rev. A **27**, 2913–2924 (1983)
398. K. Omidvar, A.M. McAllister, Evaluation of high-level bound-bound and bound-continuum hydrogenic oscillator strengths by asymptotic expansion. Phys. Rev. A **51**, 1063–1066 (1995)
399. M. Pajek, R. Schuch, Radiative recombination in the low-energy regime. Phys. Lett. A **166**, 235–237 (1992)
400. S. Schippers, D. Bernhardt, A. Müller, C. Krantz, M. Grieser, R. Repnow, A. Wolf, M. Lestin-sky, M. Hahn, O. Novotný, D.W. Savin, Dielectronic recombination of xenonlike tungsten ions. Phys. Rev. A **83**, 012711–6 (2011)
401. A. Müller, Fusion-related ionization and recombination data for tungsten ions in low to moderately high charge states. Atoms **3**, 120–161 (2015)
402. A.H. Gabriel, Some problems relating to solar line identification, in *Highlights of Astronomy*, ed. by C.R. de Jager (Dordrecht, 1971), pp. 486–494
403. S. Schippers, T. Bartsch, C. Brandau, A. Müller, G. Gwinner, G. Wissler, M. Beutelspacher, M. Grieser, A. Wolf, R.A. Phaneuf, Dielectronic recombination of lithiumlike Ni^{25+} ions: high-resolution rate coefficients and influence of external crossed electric and magnetic fields. Phys. Rev. A **62**, 022708–12 (2000)
404. V.V. Flambaum, A.A. Gribakina, G.F. Gribakin, C. Harabati, Electron recombination with multicharged ions via chaotic many-electron states. Phys. Rev. A **66**, 012713–7 (2002)
405. G.F. Gribakin, S. Sahoo, Mixing of dielectronic and multiply excited states in electronion recombination: a study of Au^{24+}. J. Phys. B: At. Mol. Opt. Phys. **36**, 3349–3370 (2003)
406. V.A. Dzuba, V.V. Flambaum, G.F. Gribakin, C. Harabati, Chaos-induced enhancement of resonant multielectron recombination in highly charged ions: statistical theory. Phys. Rev. A **86**, 022714–9 (2012)
407. V.A. Dzuba, V.V. Flambaum, G.F. Gribakin, C. Harabati, M.G. Kozlov, Electron recombination, photoionization, and scattering via many-electron compound resonances. Phys. Rev. A **88**, 062713–6 (2013)
408. I.L. Beigman, V.P. Shevelko, H. Tawara, Direct electron-impact single ionization of medium-Z ions from the ground and excited states. Phys. Scr. **53**, 534–540 (1996)
409. R.E.H. Clark, J. Abdallah Jr., Atomic data for titanium. Phys. Scr. **T37**, 28–34 (1991)
410. V.P. Shevelko, L.A. Vainshtein, E.A. Yukov, Cross sections and rate coefficients for inelastic electron collisions with carbon and oxygen ions. Phys. Scr. **T28**, 39-48 (1989); errata, Physica Scripta **44**,408 (1991)
411. P.A.Z. van Emmichoven, M.E. Bannister, D.C. Gregory, C.C. Havener, R.A. Phaneuf, E.W. Bell, X.Q. Guo, J.S. Thompson, M. Sataka, Electron-impact ionization of Si^{6+} and Si^{7+} ions. Phys. Rev. A **47**, 2888–2892 (1993)
412. M. S. Pindzola, private commun., 1993
413. D.C. Gregory, M.S. Huq, F.W. Meyer, D.R. Swenson, M. Staka, S. Chntrenne, Electron-impact ionization cross-section measurements for U^{10+}, U^{13+}, U^{16+}. Phys. Rev. A **41**, 106–115 (1990)

414. W. Lotz, Electron-impact ionization cross-sections and ionization rate coefficients for atoms and ions from hydrogen to calcium Z. Phys. **216**, 241–247 (1968)

415. M.S. Pindzola, M.J. Buie, Electron-impact ionization of uranium atomic ions. Phys. Rev. A **39**, 1029–1032 (1989)

416. M.S. Pindzola, D.C. Griffin, C. Bottcher, M.J. Buie, D.C. Gregory, Electron-impact ionization data for the nickel isonuclear sequence. Physica Scripta. **T37**, 35–46 (1991)

417. M.H. Chen, K.J. Reed, D.L. Moores, Contributions of resonant excitation double autoionization to the electron-impact ionization of Fe^{15+}. Phys. Rev. Lett. **64**, 1350–1353 (1990)

418. A. Burgess, M.C. Chidichino, Electron inpact ionization of complex ions. Mon. Not. R. Soc. Astron. Soc. **203**, 1269–1280 (1983)

419. M. Arnaud, R. Rothenflug, An updated evaluation of recombination and ionization rates. Astron. Astrophys. Suppl. Ser. **60**, 425–457 (1985)

420. T. Kato, K. Masai, M. Arnaud, Comparison of ionization rate coefficients of ions from hydrogen through nickel. NIFS-DATA-014, Nagoya, Japan (1991)

421. K.J. LaGattuta, Y. Hahn, Electron impact ionization of Fe15+ by resonant excitation double Auger ionization. Phys. Rev. A **24**, 2273–2276 (1981)

422. R.J.W. Henry, A.Z. Msezane, Cross sections for inner-shell excitation of Na-like ions. Phys. Rev. A **26**, 2545–2550 (1982)

423. A. Müller, K. Tinschert, G. Hofmann, E. Salzborn, G.H. Dunn, Resonances in electron-impact single, double, and triple ionization of heavy metal ions. Phys. Rev. Lett. **61**, 70–73 (1988)

424. A. Müller, G. Hofmann, K. Tinschert, E. Salzborn, Dielectronic capture with subsequent two-electron emission in electron-impact ionization of C^{3+} ions. Phys. Rev. Lett. **61**, 1352–1355 (1988)

425. A. Müller, K. Tinschert, C. Achenbach, E. Salzborn, R. Becker, A new technique for the measurement of ionization cross sections with crossed electron and ion beams. Nucl. Instrum. Methods Phys. Res. B **10**(11), 204–206 (1985)

426. A. Müller, Ion Formation Processes: Ionization, in *ion-electron collisions, In Physics of Ion Impact Phenomena*, ed. by D. Mathur, Springer Ser, Chem. Phys., vol. 54, (Springer, Berlin, Heidelberg, 1991), pp. 13–90

427. V.P. Shevelko, H. Tawara, F. Scheuermann, B. Fabian, A. Müller, E. Salzborn, Semiempirical formulae for electron-impact double-ionization cross sections of light positive ions. J. Phys. B: At. Mol. Opt. Phys. **38**, 525–545 (2005)

428. V.P. Shevelko, H. Tawara, IYu. Tolstikhina, F. Scheuermann, B. Fabian, A. Müller, E. Salzborn, Double Ionization of heavy positive ions by electron impact: empirical formula and fitting parameters for ionization cross sections. J. Phys. B: At. Mol. Opt. Phys. **39**, 1499–1516 (2006)

429. J.A. Syage, Electron-impact cross sections for multiple ionization of Ar: detector gain effects revealed. J. Phys. B: At. Mol. Opt. Phys. **24**, L527–L532 (1991)

430. A. Müller, K. Tinschert, C. Achenbach, R. Becker, E. Salzborn, Electron impact double ionisation of Ar+ and Ar4+ ions. J. Phys. B: At. Mol. Phys. **18**, 3011–3024 (1985)

431. M. Stenke, K. Aichele, D. Hathiramani, G. Hofmann, M. Steidl, R. Völpel, V.P. Shevelko, H. Tawara, E. Salzborn, Electron-impact multiple ionization of singly and multiply charged tungsten ions. J. Phys. B: At. Mol. Opt. Phys. **28**, 4853–4859 (1995)

432. C.D. Loch, M.S. Pindzola, N.R. Badnell, F. Scheuermann, K. Kramer, K. Huber, E. Salzborn, Electron-impact ionization of Bi^{q+} for q=1·10. Phys. Rev. A **70**, 052714–9 (2004)

433. H. Tawara, T. Kato, Total and partial ionization cross sections of atoms and ions by electron impact. At. Data Nucl. Data Tables **36**, 167–353 (1987)

434. V.P. Shevelko, H. Tawara, E. Salzborn, Mutiple-ionization cross sections of atoms and positive ions by electron impact. Report NIFS-DATA-27, National Institute for Fusion Science, Nagoya, Japan (1995)

435. V.P. Shevelko, H. Tawara, Semiempirical formulae for multiple ionization of neutral atoms and positive ions by electron impact. J. Phys. B **28**, L589–L594 (1995)

436. C. Bélenger, P. Defrance, E. Salzborn, V.P. Shevelko, H. Tawara, D.B. Uskov, Double ionization of neutral atoms, positive and negative ions by electron impact. J. Phys. B **30**, 2667–2679 (1997)

437. H. Tawara, T. Kato, Electron impact ionization data for atoms and ions. Report NIFS-DATA-51, National Institute for Fusion Science, Nagoya, Japan (1999)
438. W. Lotz, Subshell binding energies of atoms and ions from hydrogen to zinc. J. Opt. Soc. Am. **59**, 915–921 (1968); Electron Binding Energies in Free Atoms. ibid, **60**, 206–210 (1970)
439. J. Scofield, Ionization Energies, Internal Report, LLNL, CA 94550, USA
440. J.C. Halle, H.H. Lo, W. Fite, Ionization of uranium atoms bz electron impact. Phys. Rev. A **23**, 1708–1716 (1981)
441. M. Steidl, D. Hathiramani, G. Hofmann, M. Stenk,e R. Völpel, E. Salzborn, XIX ICPEAC, book of abstracts, ed. by J.B.A. Mitchell, J.W. McConkey, C.E. Brion, (Whistler, British Columbia, Canada 26 July-1 August 1995), pp. 103
442. L. House, Ionization equilibrium of the elements from H to Fe. Astrophys. J. Suppl. **8**, 307 (1964)
443. A.V. Gurevich, L.P. Pitaevskii, Recombnation coefficient in a dense low-temperature plasma. Zh. Eksp. Teor. Fiz. **46**, 1281 (1964). [Sov. Phys. JETP **19**, 870 (1964)]
444. T. Matsuo, T. Tonuma, H. Kumagai, H. Tawara, Cross sections of recoil Ne^{i+}-ion production through pure ionization, electron loss and electron capture of projectiles in 1.05-MeV/amu Ar^{q+} + Ne collisions. Phys. Rev. A **50**, 1178–1183 (1986)
445. I. Ben-Itzhak, T.G. Gray, J.L. Legg, J.H. McGuire, Inclusive and exclusive cross sections for multiple ionization by fast, highly charged ions in the independent-electron approximation. 1988. Phys. Rev. A **37**, 3685–3691 (1988)
446. T. Tonuma, H. Kumagai, T. Matsuo, H. Tawara, Coincidence measurements of slow recoil ions with projectile ions in 42-MeV Ar^{q+} - Ar collisions. Phys. Rev. A **40**, 6238–6245 (1989)
447. H. Tawara, T. Tonuma, H. Kumagai, T. Matsuo, Production of recoil Ne^{i+} ions accompanied by electron loss and capture of 1.05-MeV/amu Ne^{q+} (q = 2, 4, 6, 8, and 10) ions. Phys. Rev. A **41**, 116–122 (1990)
448. C.L. Cocke, R.E. Olson, Recoil ions. Phys. Rep. **205**, 153–219 (1991)
449. K. Kelbch, J. Ullrich, R. Mann, P. Richard, H. Schmidt-Böking, Cross sections for the production of highly charged argon and xenon recoil ions in collisions with high-velocity uranium projectiles. J. Phys. **18**, 323–336 (1985)
450. K.S. Kelbch, H. Schmidt-Böcking, J. Ullrich, R. Schuch, E. Justiniano, M. Ingwersen, C.L. Cocke, The contribution of K-electron capture for the production of highly charged Ne recoil ions by 156 MeV bromine impact. Z. Phys. A **317**, 9–14 (1984)
451. A. Müller, R. Schuch, W. Groh, E. Salzborn, H.F. Beyer, P.H. Mokler, R.E. Olson, Multiple-electron capture and ionization in heavy-ion-atom collisions. Nucl. Instrum. Methods B **2425**, 111–114 (1987)
452. H. Berg, R. Dörner, C. Kelbch, S. Kelbch, J. Ullrich, S. Hagmann, P. Richard, H. Schmidt-Böcking, Multiple ionisation of rare gases by high-energy uranium ions. J. Phys. B **21**, 3929–3939 (1988)
453. S. Kelbch, C.L. Cocke, S. Hagmann, M. Horbatsch, C. Kelbch, R. Koch, H. Schmidt-Böcking, J. Ullrich, Recoil-ion production cross sections and differential scattering angle dependences in 2.5-15 MeV F^{n+} (n = 4, 6, 8) on Ne collisions. J. Phys. B **23**, 1277–1301 (1990)
454. W. Wu, S. Datz, N.L. Jones, H.F. Krause, B. Rosner, K.D. Sorge, C.R. Vane, Double ionization of He by fast protons at large energy transfer. Phys. Rev. Lett. **76**, 4324–4327 (1996)
455. H.E. Berg, Die vielfashionization von edelgasen in hochenergetischen schwerionenstössen. Ph.D. thesis (in German) Report GSI-**93-12** (Darmstadt, 1993)
456. J.H. McGuire, L. Weaver, Independent electron approximation for atomic scattering by heavy particles. Phys. Rev. A **16**, 41–47 (1977)
457. M. Horbatsch, Semiclassical description of multiple electron capture and ionization in fast bare nucleus–rare gas collisions. Z. Phys. D **1**, 337–345 (1986)
458. M. Horbatsch, Calculation of transfer ionisation processes in ion-atom collisions. J. Phys. B **19**, L193–L198 (1986)
459. M. Horbatsch, R.M. Dreizler, Semiclassical description of multiple electron capture and ionization in fast bare nucleus-rare gas collisions. Z. Phys. D **2**, 183–191 (1986)

460. M. Horbatsch, Theory of multiple ionization and capture in energetic ion-atom collisions. Z. Phys. D **21**, S63–S67 (1991)

461. M. Horbatsch, Independent particle model description of multiple ionization dynamics in fast ion-atom collisions. J. Phys. B **25**, 3797–3821 (1992)

462. T. Kirchner, H. Tawara, I.Y. Tolstikhina, A.D. Ulantsev, V.P. Shevelko, Th Stöhlker, Multi-electron ionization of atoms by fast ions: an approximation by normalized exponentials. Tech. Phys. **51**, 1127–1136 (2006). [Zh. Tech. Fyz. Vol. **76**, 22–30 (2006)]

463. IYu. Tolstikhina, V.P. Shevelko, Multiple ionization of atoms by highly charged ions. Phys. Scr. **90**, 074033–6 (2015)

464. H.O. Heber, R.L. Watson, G. Sampoll, B.B. Bandong, Three-shell model for independent-electron processes in heavy-ionatom collisions. Phys. Rev. A **42**, 6466–6470 (1990)

465. T. Kirchner, L. Gulyas, H.J. Ludde, E. Engel, R.M. Dreizler, Influence of electronic exchange on single and multiple processes in collisions between bare ions and noble-gas atoms. Phys. Rev. A **58**, 2063–2076 (1998)

466. I. Lesteven-Vaisse, D. Hennecart, R. Gayet, Multiple ionization cross-sections of rare gas atoms by impact of highly charged particles at 35 MeV/a.m.u. J. Phys. (France) **49**, 1529–1544 (1988)

467. R. Gayet, Multiple capture and ionization in high-energy ion-atom collisions. J. Phys. (France) **50**, C1-53–C1-70 (1989)

468. A. Salin, Helium ionisation by high-energy ions at a function of impact parameter and projectile scattering angle. J. Phys. B **22**, 3901–3914 (1989)

469. V. Krishnamurthi, I. Ben-Itzhak, K.G. Carnes, Projectile charge dependence of ionization and fragmentation of CO in fast collisions. J. Phys. B **29**, 287–297 (1996)

470. R.E. Olson, A. Salop, Charge-transfer and impact-ionization cross sections for fully and partially stripped positive ions colliding with atomic hydrogen. Phys. Rev. A **16**, 531–541 (1977)

471. R.E. Olson, Ion-atom differential cross sections at intermediate energies. Phys. Rev. A **27**, 1871–1878 (1983)

472. J. Ullrich, M. Horbatsch, V. Dangendorf, S. Kelbch, H. Schmidt-Böcking, Scattering-angle-dependent multiple ionisation cross sections in high-energy heavy-ion-atom collisions. J. Phys. B **21**, 611–624 (1988)

473. R.D. DuBois, S.T. Manson, Multiple-ionization channels in proton-atom collisions. Phys. Rev. A **35**, 2007–2025 (1987)

474. C.J. Patton, M.B. Shah, M.A. Bolorizadeh, J. Geddes, H.B. Gilbody, Ionization in collisions of fast H+ and He^{2+} ions with Fe and Cu atoms. J. Phys. B **28**, 3889–3899 (1995)

475. B. Sulik, G. Hock, D. Berényi, Charge scaling of ionisation probabilities in ion-atom collisions for zero impact parameter. J. Phys. B: At. Mol. Phys. **17**, 3239–3244 (1984)

476. I. Kadar, S. Ricz, V.A. Shchegolev, B. Sulik, D. Varga, J. Vegh, D. Berenyi, G. Hock, Auger electron spectra in 5.5 MeV amu-q Neq+ and Ar^{q+} ion impact on Ne. J. Phys. B: At. Mol. Phys. **18**, 275–287 (1985)

477. B. Sulik, I. Kádár, S. Ricz, D. Varga, J. Vegh, G. Hock, D. Berényi, A simple theoretical approach to multiple ionization and its application for 5.1 and 5.5 MeV/u X^{q+} + Ne collisions. Nucl. Instrum. Methods B **28**, 509–518 (1987)

478. T. Mukoyama, S. Ito, B. Sulik, G. Hock, Wave function effect on the ionization probability in the geometrical model. Bull. Inst. Chem. Res. Kyoto Univ. **68**, 281–294 (1991)

479. B. Sulik, K. Tőkési, Y. Awaya, T. Kambara, Y. Kanai, Single and double K-shell vacancy production in N^{7+} + Ti collisions. Nucl. Instrum. Methods B **154**, 286–290 (1999)

480. A.D. Ulantsev, Geometrical model: the single and multiple ionization of ions and atoms in ion-atom collisions. J. Phys. B: At. Mol. Opt. Phys. **41**, 165203–9 (2008)

481. R.L. Kauffman, C.W. Woods, K.A. Jamison, P. Richard, Relative multiple ionization cross sections of neon by projectiles in the 1-2-MeV/amu energy range. Phys. Rev. A **11**, 872–883 (1975)

482. F. Folkmann, R. Mann, H.F. Beyer, Excited states of few-electron recoil ions from ion impact on neon studied by x-ray and electron measurements. Phys. Scripta **T3**, 88–95 (1983)

483. D. Schneider, M. Prost, R. DuBois, N. Stolterfoht, Ne K-Auger electron emission following high-energy Ne^{9+} and Ar^{9+} ion impact on Ne. Phys. Rev. A **25**, 3102–3107 (1982)

484. I. Kadar, S. Ricz, V.A. Shchegolev, D. Varga, J. Vegh, D. Berenyi, G. Hock, B. Sulik, Characterization of multiple ionization processes by means of auger spectra measured in 5.5 MeV/u Ne^{3+}, Ne^{10+} and Ar^{6+} - Ne collisions. Phys. Lett. A **115**, 439–442 (1986)

485. G.H. McGuire, P. Richard, Procedure for computing cross sections for single and multiple ionization of atoms in the binary-encounter approximation by the impact of heavy charged particles. Phys. Rev. A **8**, 1374–1384 (1973)

486. I.C. Percival, D. Richards, Classical theory of transitions between degenerate states of excited hydrogen atoms in plasmas. J. Phys. B **12**, 2051–2065 (1979)

487. E.G. Cavalcanti, G.M. Sigaud, E.C. Montenegro, M.M. SantÁnna, H. Schmit-Böcking, Postcollisional effects in multiple ionization of neon by protons. J. Phys. B **35**, 3937–3944 (2002)

488. T.A. Carlson, W.E. Hunt, M.O. Krause, Relative abundances of ions formed as the result of inner-shell vacancies in atoms. Phys. Rev. **151**, 41–47 (1966)

489. H. Poth, Electron cooling: Theory, experiment, application. Phys. Rep. **196**, 135–197 (1990)

490. R.D. DuBois, O.de Lucio, M. Thomason, G. Weber, Th. Stöhlker, K. Beckert, P. Beller, F. Bosch, C. Brandau, A. Gumberidze, S. Hagmann, C. Kozhuharov, F. Nolden, R. Reuschl, J. Razdkjewicz, P. Spiller, U. Spillmann, M. Steck, S. Trotsenko, Beam lifetimes for low-charge-state heavy ions in the GSI storage rings. Nucl. Instrum. Methods B **261**, 230–233 (2007)

491. A. Krämer, et al., in Proc. of the 8th European Particle Accelerator Conf., EPAC 2002, Paris, 3 - 7 June 2002 (Geneva: EPS-IGA, CERN, 2002) p. 2547 (2002); http://accelconf.web.cern.ch/accelconf/e02/PAPERS/WEPLE116.pdf

492. P. Spiller, *GSI Scientific Report 2010* (GSI, Darmstadt, 2011), p. 270

493. H.J. Kunze, S.S. Ellwi, Ž. Andreić, X-ray lasing in ablative capillary discharges. Czech. J. Phy. **56**(S2), B280–B290 (2006)

494. Ž. Andreić, H.J. Kunze, I. Tolstikhina, Evidence of lasing on the Balmer-α line of OVIII in an ablative capillary discharge. Opt. Lett. **40**(11), 2600–2602 (2015)

495. H.J. Kunze, S. Ellwi, Ž. Andreić, Lasing in an ablative capillary discharge with structured return conductor. Phy. Lett. A **334**(1), 37–41 (2005)

496. D.L. Matthews, P.L. Hagelstein, M.D. Rosen, M.J. Eckart, N.M. Ceglio, A.U. Hazi, H. Medecki, B.J. MacGowan, J.E. Trebes, B.L. Whitten, E.M. Campbell, C.W. Hatcher, A.M. Hawryluk, R.L. Kauffman, L.D. Pleasance, G. Rambach, J.H. Scofield, G. Stone, T.A. Weaver, Demonstration of a soft x-ray amplifier. Phys. Rev. Lett. **54**, 110–113 (1985)

497. S. Suckewer, C.H. Skinner, H. Milchberg, C. Keane, D. Voorhees, Amplification of stimulated soft X-ray emission in a confined plasma column. Phys. Rev. Lett. **55**, 1753–1756 (1985)

498. J. Zhang, X-Ray Lasers 2004. Institute of Physics Conference Series **186**. CRC Press (2005)

499. H. Daido, Review of soft X-ray laser researches and developments. Rep. Progr. Phy. **65**(10), 1513–1576 (2002)

500. H.J. Shin, D.E. Kim, T.N. Lee, Soft-X-ray amplification in a capillary discharge. Phys. Rev. E **50**, 1376–1382 (1994)

501. L. Aschke, Y.V. Ralchenko, E.A. Solov'ev, H.J. Kunze, Enhanced Balmer-alpha emission in interpenetrating plasmas. Le Journal de Physique IV **11**(Pr2), Pr2547–Pr2550 (2001)

502. https://en.wikipedia.org/wiki/Periodic_table

503. Yu. Oganessian, Superheavy elements. Pure Appl. Chem. **76**, 1715–1734 (2004)

504. Ch.E. Düllmann, R.-D. Herzberg, N. Nazarewicz, Y. Oganessin, (eds.), *Special issue on superheavy elements*. Nuclear Physics A, **944** (December 2015) pp. 1–690 (2015)

505. K. Morita, A. Yoshida, T.T. Inamura, M. Koizumi, T. Nomura, M. Fujioka, T. Shinozuka, H. Miyatake, K. Sueki, H. Kudo, Y. Nagai, T. Toriyama, K. Yoshimura, Y. Hatsukawa, Nucl. Instrum. Methods B **70**, 220–226 (1992)

506. M. Leino, In-flight separation with gas-filled systems. Nucl. Instrum. Methods B **126**, 320–328 (1997)

507. Y.T.V. Oganessian, K. Utyonkov, Y.V. Lobanov, F.S. Abdullin, A.N. Polyakov, I.V. Shirokovsky, A.N. YuS Tsyganov, S. Mezentsev, V.G. Iliev, A.M. Subbotin, G.V. Sukhov, K. Subotic Buklanov, K.J. YuA Lazarev, J.F. Moody, N.J. Wild, M.A. Stoyer, R.W. Stoyer, C.A. Laue Lougheed, Average charge states of heavy atoms in dilute hydrogen. Phys. Rev. C **64**, 06409–6 (2001)

508. K.E. Gregorich, W. Loveland, D. Peterson, P.M. Zielinski, S.L. Nelson, Y.H. Chung, C.M. ChE Düllmann, K. Folden, R. Aleklett, D.C. Eichler, J.P. Hoffman, G.K. Omtvedt, J.M. Pang, S. Schwantes, P. Soverna, R. Sprunger, R.E. Sudowe, H. Nitsche, Wilson, Attempt to confirm superheavy element production in the ^{48}Ca + ^{238}U reaction. Phys. Rev. C **72**, 014605–7 (2005)

509. J. Khuyagbaatar, D. Ackermann, L.-L. Andersson, J. Ballof, W. Brüchle, C.E. Düllmann, J. Dvorak, K. Eberhardt, K. Even, A. Gorshkov, R. Graeger, F.-P. Hessberger, D. Hild, R. Hoischen, E. Jäger, B. Kindler, J.V. Kratz, S. Lahiri, B. Lommel, M. Maiti, E. Merchan, D. Rudolph, M. Schädel, H. Schaffner, B. Schausten, E. Schimpf, A. Semchenkov, A. Serov, A. Türler, A. Yakushev, Nucl. Instrum. Methods A **689**, 40–46 (2012)

510. theory and experiment, J. Khuyagbaatar, V.P. Shevelko, A. Borschevsky, C.E. Düllmann, IYu. Tolstikhina, A. Yakushev, Average charge states of heavy and superheavy ions passing through a rarified gas. Phys. Rev. A **88**, 042703–8 (2013)

511. J. Khuyagbaatar, A. Yakushev, D. ChE Düllmann, L.-L. Ackermann, M. Andersson, M. Asai, R.A. Block, H. Boll, D.M. Brand, M. Cox, X. Dasgupta, A. Di Derkx, K. Nitto, J. Eberhardt, M. Even, C. Evers, U. Fahlander, J.M. Forsberg, N. Gates, P. Gharibyan, K.E. Golubev, J.H. Gregorich, W. Hamilton, R.-D. Hartmann, F.P. Herzberg, D.J. Hessberger, J. Hinde, R. Hoffmann, A. Hollinger, E. Hübner, B. Jäger, J.V. Kindler, J. Kratz, N. Krier, M. Kurz, S. Laatiaoui, R. Lahiri, B. Lang, M. Lommel, K. Maiti, S. Miernik, A. Minami, C. Mistry, H. Mokry, J.P. Nitsche, G.K. Omtvedt, P. Papadakis Pang, D. Renisch, J. Roberto, D. Rudolph, J. Runke, K.P. Rykaczewski, L.F. Sarmiento, M. Schädel, B. Schausten, A. Semchenkov, D.A. Shaughnessy, P. Steinegger, J. Steiner, E.E. Tereshatov, P. Thörle-Pospiech, K. Tinschert, T. Torres De Heidenreich, N. Trautmann, A. Türler, J. Uusitalo, D.E. Ward, M. Wegrzecki, N. Wiehl, S.M. Van Cleve, V. Yakusheva, ^{48}Ca + ^{249}Bk fusion reaction leading to element Z = 117: long-lived α-Decaying ^{270}Db and discovery of ^{266}Lr. Phys. Rev. Lett. **112**, 172501–5 (2014)

512. M. Toulemonde, E. Paumier, C. Dufour, Thermal spike model in the electronic stopping power regime. Radiat. Eff. Defects Solids **126**, 201–206 (1993)

513. S.T. de Zwart, A.G. Drentje, A.L. Boers, R. Morgenstern, Electron emission induced by multiply charged Ar ions impinging on a tungsten surface. Surf. Sci. **217**, 298–316 (1989)

514. G. Schiwietz, G. Xiao, P.L. Grande, E. Luderer, R. Pazirandeh, U. Stettner, Determination of the electron temperature in the thermal spike of amorphous carbon. Europhys. Lett. **47**, 384–390 (1999)

515. R.M. Papaléo, M.R. Silva, R. Leal, P.L. Grande, M. Roth, B. Schattat, G. Schiwietz, Direct evidence for projectile charge-state dependent crater formation due to fast ions. Phys. Rev. Lett. **101**, 167601–4 (2008)

516. S. Gupta, H.G. Gehrke, J. Krauser, C. Trautmann, D. Severin, M. Bender, H. Rothard, H. Hofsäss, Conduction ion tracks generated by charge-selected swift heavy ions. Nucl. Instr. Meth. B **381**, 76–83 (2016)

517. N. Matsunami, M. Kato, M. Sataka, S. Okayasu, Disordering of ultra thin WO3 films by high-energy ions. Nucl. Instr. Meth. B **409**, 272–276 (2017)

518. G. Gamov, Zur Quantentheorie des Atomkernes. Z. Phys. **51**, 204–212 (1928)

519. H. Geiger, J.M. Nuttall, The ranges of a particles from uranium. Philos. Magazine Ser. 6(23), 439–445 (1912)

520. E. Fermi, *A Course Given by Enrico Fermi at the University of Chicago. Nuclear Physics.* University of Chicago Press, ISBN 978-0226243658 (1950)

521. K. Takahashi, K. Yokoi, Nuclear decays of highly ionized heavy atoms in stellar interiors. Nucl. Phys. A **404**, 578–598 (1983)

522. J.N. Bahcall, Theory of bound-state beta decay. Phys. Rev. **124**, 495–499 (1961)

523. A.G.W. Cameron, Photobeta reactions in stellar interiors. Aph. J. **130**, 452 (1959)

524. K. Takahashi, K. Yokoi, Beta-decay rates of highly ionized heavy atoms in stellar interiors. At. Data Nucl. Data Tables **36**, 375–409 (1987)

525. Yu.A. Litvinov, F. Bosch, H. Geissel, J. Kurcewicz, Z. Patyk, N. Winckler, L. Batist, K. Beckert, D. Boutin, C. Brandau, L. Chen, C. Dimopoupou, B. Fabian, T. Faestermann, A. Fragner, L. Grigerenko, E. Haettner, S. Hess, P. Kienle, R. Knöbel, C. Kozhuharov, S.A. Litvinov, L. Maier, M. Mazzocco, F. Montes, G. Münzenberg, A. Mizimarra, C. Nociforo, F. Noiden, M. Pfützner, W.R. Plass, A. Prochazka, R. Reda, R. Reuschl, C. Scheidenberger, M. Steck, T. Stöhlker, S. Torilov, M. TRassinelli, B. Sun, H. Weick, N. Winckler, Measurement of the + and Orbital Electron-Capture Decay Rates in Fully Ionized, Hydrogenlike, and Heliumlike ^{140}Pr Ions. Phys. Rev. Lett. **99**, 262501 (2007)

526. N. Winckler, H. Geissel, Yu.A. Litvinov, K. Beckert, F. Bosch, D. Boutin, C. Brandau, L. Chen, C. Dimopoulou, H.G. Essel, B. Fabian, T. Faestermann, A. Fragner, E. Haettner, S. Hess, P. Kienle, R. Knöbel, C. Kozhuharov, S.A. Litvinov, M. Mazzocco, M. Montes, Orbital electron capture decay of hydrogen- and helium-like ^{142}Pm ions. Phys. Lett. B **679**, 36–40 (2009)

527. J.K. Tuli, Nuclear data sheets for A = 142. Nucl. Data Sheets **89**, 641–796 (2000)

528. N. Winckler, Nuclear orbital electron capture of stored highly-ionized ^{140}Pr and ^{142}Pm ions. Ph.D. thesis, Giessen University (2009)

529. Z. Patyk, J. Kurcewicz, F. Bosch, H. Geissel, Y.A. Litvinov, M. Pfützner, Orbital electron capture decay of hydrogen- and helium-like ions. Phys. Rev. C **77**, 014306–4 (2008)

530. G. Strang, *Introduction to Linear Algebra* 3rd ed. Wellesley-Cambridge Press (1998)

531. T.D. Lee, C.N. Yang, Question of parity conservation in weak interactions. Phys Rev. **104**, 254–258 (1956); Erratum Phys. Rev. **106**, 1371 (1957)

532. C.S. Wu, E. Ambler, R.W. Hayward, D.D. Hoppes, R.P. Hudson, Experimental test of parity conservation in beta decay. Phys. Rev. **105**, 1413–1415 (1957)

533. L. Grodzins, M. Goldhaber, A.W. Sunyar, Helicity of neutrinos. Phys. Rev. **109**, 1015–1017 (1958)

534. R.P. Feynman, M. Gell-Mann, Theory of the fermi interaction. Phys. Rev. **109**, 193–198 (1958)

535. Patyk, Z.: Private communication (2009)

536. I.N. Borzov, E.E. Saperstein, S.V. Tolokonnikov, Magnetic moments of spherical nuclei: status of the problem and unsolved issues. Phys. Atomic Nuclei **71**, 469–491 (2008)

Index

Printed in the United States
By Bookmasters